International Conference on Materials Science and Manufacturing Engineering (MSME 2018)

MATEC Web of Conferences Volume 253 (2019)

Paris, France
8 - 10 November 2018

Editor:

David Bassir

ISBN: 978-1-5108-7904-1

Printed from e-media with permission by:

Curran Associates, Inc.
57 Morehouse Lane
Red Hook, NY 12571

Some format issues inherent in the e-media version may also appear in this print version.

Printed by Curran Associates, Inc. (2019)

For additional information, please contact EDP Sciences – Web of Conferences at the address below.

EDP Sciences – Web of Conferences
17, Avenue du Hoggar
Parc d'Activité de Courtabœuf
BP 112
F-91944 Les Ulis Cedex A
France

Phone: +33 (0) 1 69 18 75 75
Fax: +33 (0) 1 69 28 84 91

contact-edps@webofconferences.org

Additional copies of this publication are available from:

Curran Associates, Inc.
57 Morehouse Lane
Red Hook, NY 12571 USA
Phone: 845-758-0400
Fax: 845-758-2633
Email: curran@proceedings.com
Web: www.proceedings.com

TABLE OF CONTENTS

PREFACE

MSME 2018 was held in Paris during November 8-10, 2018. MSME 2018 is organized by the Hong Kong Society of Mechanical Engineers (HKSME) and supported by the Sino-French Association of Sciences and Technologies (AFCST), the Association ASMDO and international partner of publication EDP Sciences. The conference provides a useful and wide platform both for display the latest research and for exchange of research results and thoughts in broad Materials Science and Manufacturing Engineering. The participants of the conference were from almost every part of the world, with background of either academia or industry, even well-known enterprise. The success and prosperity of the conference is reflected high level of the papers received.

The proceedings are a compilation of the accepted papers and represent an interesting outcome of the conference. This book covers 3 chapters: Material Properties & Characterization; Material Design & Modelling; Material Manufacturing & Process.

We would like to acknowledge all of those who supported MSME 2018. Each individual and institutional help were very important for the success of this conference. Especially we would like to thank the organizing committee for their valuable advices in the organization and helpful peer review of the papers.

We sincerely hope that MSME 2018 will be a forum for excellent discussions that will put forward new ideas and promote collaborative researches. We are sure that the proceedings will serve as an important research source of references and the knowledge, which will lead to not only scientific and engineering progress but also other new products and processes.

Prof. David Bassir, ENS Cachan (CMLA) University of Paris Sarclay, France

COMMITTEES

Conference Co-Chairs

Prof. David Bassir, ENS Cachan (CMLA) university of Paris sarclay, France

Prof. Stefan Dimov, University of Birmingham, UK

Program Co-Chairs

Prof. Jan Taler, Cracow University of Technology, Poland

Prof. Ziad MOUMNI, ENSTA ParisTech, France

Prof. Rachid bennacer, School Ecole Normale Superieure (Cachan), France

Prof. Hamid Assadi, Brunel University London, UK

Prof. Jüri Majak, Tallinn University of Technology, Estonia

International Technical Commmitte

Assoc. Prof. Ţălu Ştefan, The Technical University of Cluj-Napoca, Romania

Dr. Han Zhang, Queen Mary University of London, UK

Dr. Dmitry Ivanov, University of Bristol, UK

Dr. Haiming Wen, Missouri University of Science and Technology, USA

Dr. Abhijit Pramanick, City University of Hong Kong, Hong Kong

edp sciences

MATEC Web of Conferences

Statement of Peer review

In submitting conference proceedings to MATEC Web of Conferences, I certify to the Publisher that I adhere to the **Policy on Publishing Integrity** of the journal in order to safeguard good scientific practice in publishing.

1. All articles have been subjected to peer review administered by the proceedings editors.
2. Reviews have been conducted by expert referees, who have been requested to provide unbiased and constructive comments aimed, whenever possible, at improving the work.
3. Proceedings editors have taken all reasonable steps to ensure the quality of the materials they publish and their decision to accept or reject a paper for publication has been based only on the merits of the work and the relevance to the journal.

Title, date and place of the conference

2018 International Conference on Materials Science and Manufacturing Engineering (MSME 2018)

Nov.8-10 2018 Paris, France

Proceedings editor(s):

Prof. David Bassir, ENS Cachan (CMLA) University of Paris Sarclay, France

Date and editor's signature

2018/11/28

MATEC Web of Conferences **253**, 01001 (2019) https://doi.org/10.1051/matecconf/201925301001
MSME 2018

Stiffness Analysis and Experimental Verification of Segment Frame of Continuous Casting Machine

Zhufeng Lei [1,a], Wenbin Su [1], Bingrui Hou[1] and Qi Gao[2]

[1] *Xi'an Jiaotong University, School of Mechanical Engineering, Xi'an, Shaanxi 710049, P.R. China.*
[2] *China National Heavy Machinery Research Institute Co.,Ltd, Xi'an, Shaanxi 710016, P.R. China.*

Abstract. To verify the stiffness, the natural vibration and the reliability of the continuous casting machine in the configuration of dynamic soft reduction. Firstly, the three-dimensional model of the segment frame was established and the finite element simulation of the structure was carried out. Then, the deviation of the roll gap in the working process of the caster was measured by analog loading test. The results show that the stiffness of the segment frame of the continuous casting machine meets the requirements. Using dynamic soft reduction can improve the centerline segregation of casting slab in the process of continuous casting, which is of important practical significance in the production practice.

1 Introduction

Since the 70s of the 20th century, with the improvement of the quality requirements of the steel slabs of various countries, the center segregation of the slab has become the critical attention of all countries. Therefore, in order to solve the slab segregation problem, dynamic soft reduction technology is widely used in continuous casting production process [1-3]. At present, SMS Siemag, Siemens VAI and Danieli Davy, which have developed relatively mature dynamic soft reduction equipment and process technology and are successfully applied to the steel industry, Distington). The companies mentioned above have developed a configuration of soft caster fan section equipment and related technologies.

The segregation of high-strength steel series in the CCC pilot production center is prominent due to the inability to meet the requirements of stable and continuous production process quality because the continuous casting machine segments do not have the functions of remote adjustment of roll gap and un-configured dynamic soft reduction [1-2]. In order to minimize centerline segregation and improve the microstructural homogeneity of the strand in subsequent rolling, technologies such as electromagnetic stirring [4-5], thermal soft reduction [6-9], dynamic soft reduction [10-17] and mechanical forging [18] have been applied in the continuous casting of steel. Among all methods, dynamic soft reduction is considered the most flexible and effective way to minimize segregation of cast steel. The dynamic soft reduction technique can be applied to minimize the centerline segregation and porosity by reducing the strand near the final solidification region to compensate for solidification shrinkage [19]. However, there is a higher requirement on the rigidity of caster frame with dynamic soft reduction.

In order to verify whether the rigidity and dynamic characteristics of the caster segment frame meet the requirements and ensure the service life of the equipment after the configuration of dynamic soft reduction. In this paper, the finite element simulation of the structural rigidity and dynamic

[a] Corresponding author: leizhufeng@qq.com

characteristics of the segment frame of a continuous casting machine was carried out. Then, hydraulic jacks were used to simulate the pressure load of the slab casting during continuous casting process to detect the actual segment stiffness of the frame characteristics.

2 The establishment of segment frame model

2.1 The establishment of three-dimensional solid model of segment frame

The three-dimensional model of the continuous casting machine segment frame was established and the finite element simulation analysis was carried out, as shown in Figure.1.

Figure 1. Segment frame three-dimensional model **Figure 2**. Simulation of slab contact with the nip roll

In order to simulate the reaction force of the slab against the segment frame in the continuous casting process and to simulate the contact between the slab and the segment nip roller, a concave unit was established to minimize the contact between the slab and the nip roll. The simulated slab contact with the nip roll is shown in Figure.2.

2.2 The establishment of finite element model of segment frame

When the finite element model of the continuous casting machine fan frame was established, the material selection is Q345B, the density $\rho = 7850kg/m^3$, the elastic modulus E = 2GPa and the Poisson's ratio $\mu = 0.3$. The total mass of the frame is 25364*kg*. Due to the meshing process, the hexahedral mesh is more likely to converge than the tetrahedral mesh and the discretization error is small and the precision is high. Therefore, in the design simulation module, the solid model is automatically meshed using a hexahedral mesh to obtain a finite element model, with a total of 977760 nodes and 442188 finite elements, as shown in Figure.3.

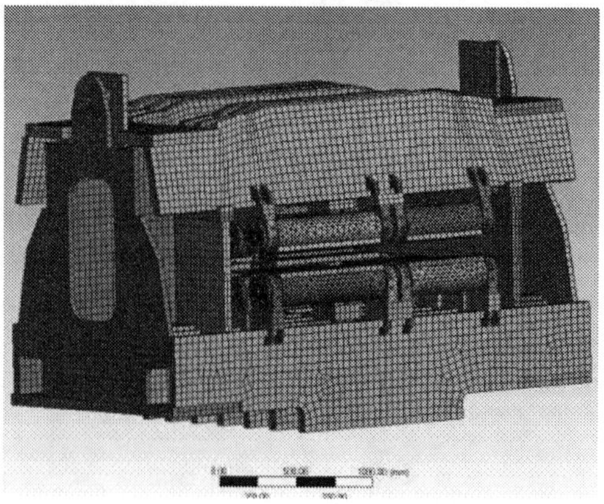

Figure 3. Segment frame finite element model

3 Finite element analysis of segment frame

3.1 Static analysis of segment frame

When the segment is working, the sector frame is mainly affected by the reaction force of the slab against the frame, the gravity of the machine frame, the clamping force of the upper and lower clamps, the bulging force, the resistance of the light pressure and the rotation resistance of the roller. Therefore, for the moment, the force loads of the segment frame are:

(1) The weight of the base itself.

(2) The reaction to the segment frame resulting from the compression of the slab.

(3) The clamping force of the upper and lower clamps.

(4) The bulging force.

(5) Resistance of the light pressure and the rotation resistance of the roller.

According to mechanical analysis, during the process of passing through the slab liquid core, the rollers of the inner and outer arc frames are deformed by the pressure, and the deformation of the outer arc frame due to the complete restraint of the outer arc frame is small. In this case, the inner arc frame can be regarded as a simply supported beam model subjected to uniform force, as shown in Figure.4.

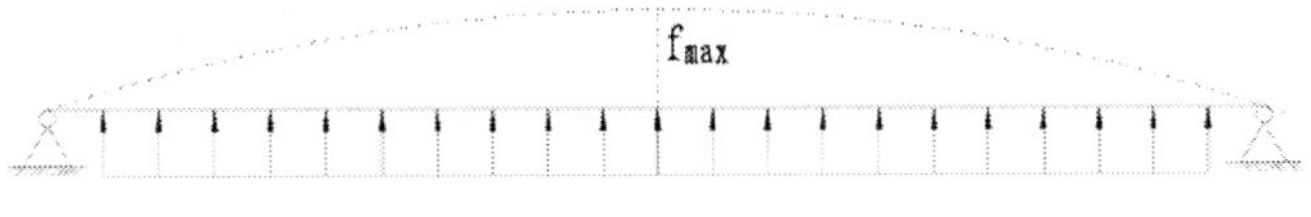

Figure 4. Simply supported beam load deformation diagram

According to mechanical analysis of the mechanics of simply supported beam, under the uniform load, the simply supported beam is deformed [20]. According to the deflection equation:

$$f_{max} = \frac{5ql^4}{384EI} \tag{1}$$

In which: f_{max} is the maximum deflection; q is uniform load; l is simply supported beam length; E is modulus of elasticity; I is Sectional moment of inertia.

It can be seen that the maximum deformation occurs at the midpoint of the beam. The maximum deformation of the segment frame occurs at the midpoint of the nip roller connection at both ends of the inner arc frame.

The segment frame is mainly affected by the reaction force of the slab against the frame, assuming that it is evenly distributed along the axial direction of the nip roller. It can be seen from Figure.4 that the force of each row of nip rollers is 300kN and the deformation of the whole segment frame is less than 1.8mm. According to equation:

$$P = \frac{W}{S} = \frac{75000}{0.082^2\,\pi/4} \tag{2}$$

The hydraulic test system needs to provide a maximum pressure of 14.8MPa; there are 2 loads applied to the stand for the FEA:

(1) Slab pressure on the frame.

(2) The gravity of the base itself.

According to the analysis results, the equivalent displacement cloud frame of the segment frame is shown in Figure 5. The maximum deformation of the frame appears at the nip roller connecting ends of the inner arc frame with the maximum deformation of 1.8718mm. The maximum deformation of the segment frame is shown in Figure 6.

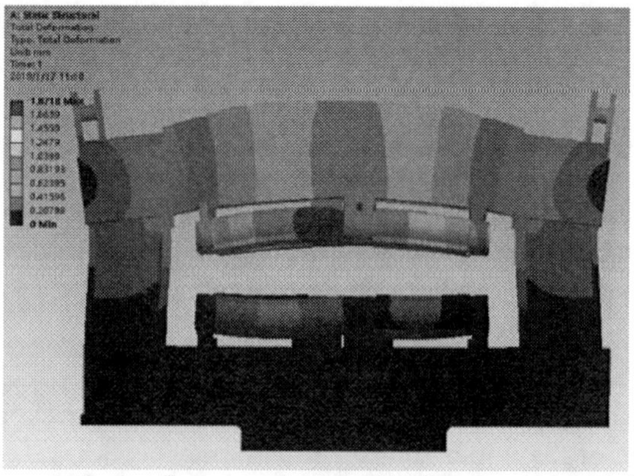

Figure 5. Sector frame equivalent displacement cloud chart

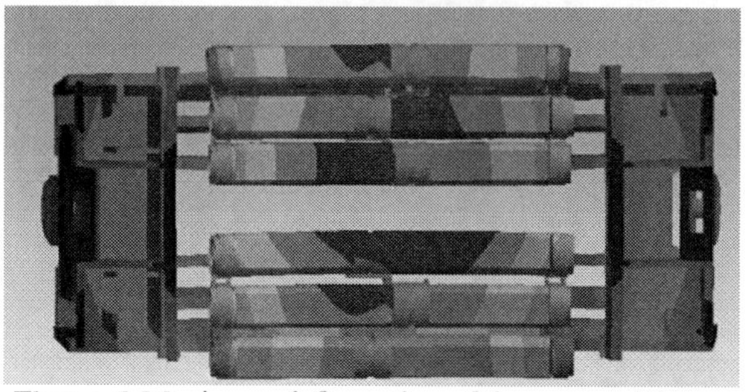

Figure 6. Maximum deformation of the segment frame

According to the theoretical analysis, the deformation of the segment frame occurs mainly in the connecting part of the nip roller of the inner arc frame, and the simulation results are in consistent with the theoretical analysis.

3.2 Dynamic analysis of segment frame

During the process of continuous casting, the slab liquid core passes through the segment frame double row nip rollers and is casted under the pressure of driving rollers. During the process of continuous casting, the slab passes through the segment frame while being subjected to great pressure of nip rollers. This process inevitably generates resonance. When the vibration frequency reaches the resonance frequency of the segment frame, the device generates resonance. Therefore, the dynamic analysis of the segment frame provides the basis for the structure anti-vibration and vibration reduction and the improvement of the overall structural reliability and service life.

The boundary conditions of the segment frame are loaded to solve the first six natural frequencies and modes of the segment frame under the pre-stress. The results are shown in Table1, the mode shapes are shown in Figure7 ~ Figure12.

Table 1. The first six natural frequencies and mode shapes analysis of segment frame

	Frequency/Hz	Mode shape
1st order	47.384	The swing of inner arc frame swing in the X direction
2nd order	55.585	The second-orderly swing of inner arc frame in the X direction
3rd order	61.628	The torsion of inner arc frame in the Y direction
4th order	85.724	The swing of outer arc frame in the X direction
5th order	97.263	The second-orderly swing of outer arc frame in the X direction
6th order	100.46	The torsion of inner arc frame in the X direction

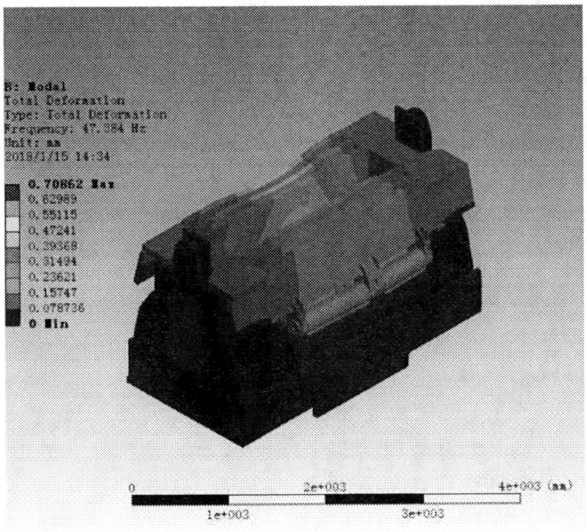

Figure 7. 1st order mode shape cloud chart

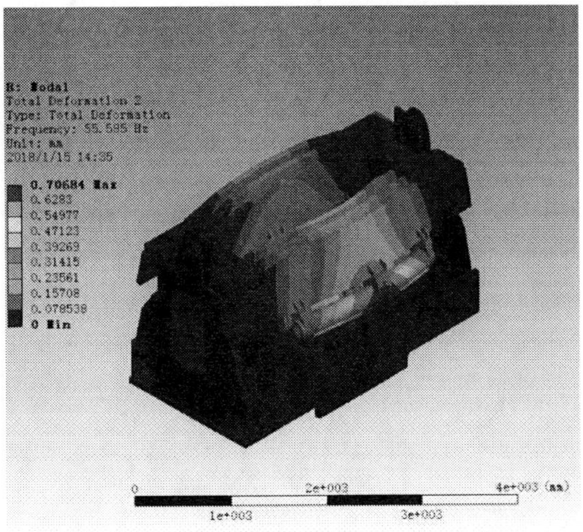

Figure 8. 2nd order mode shape cloud chart

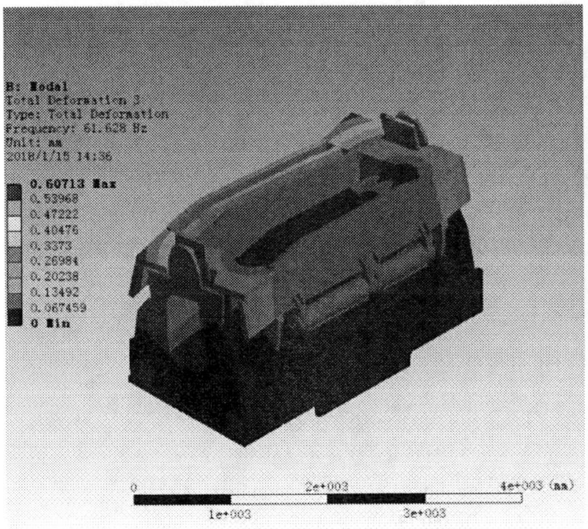

Figure 9. 3rd order mode shape cloud chart

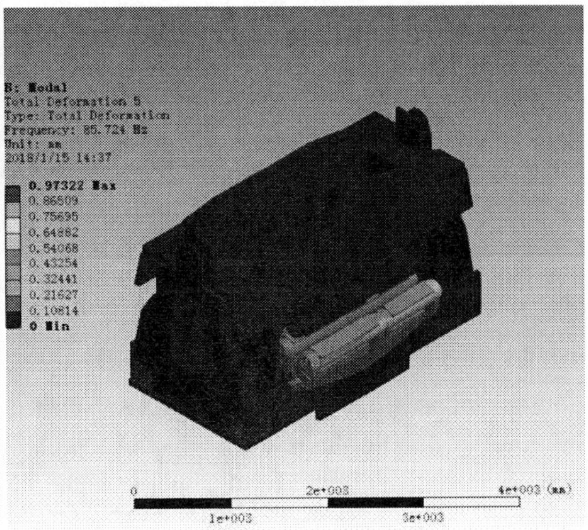

Figure 10. 4th order mode shape cloud chart

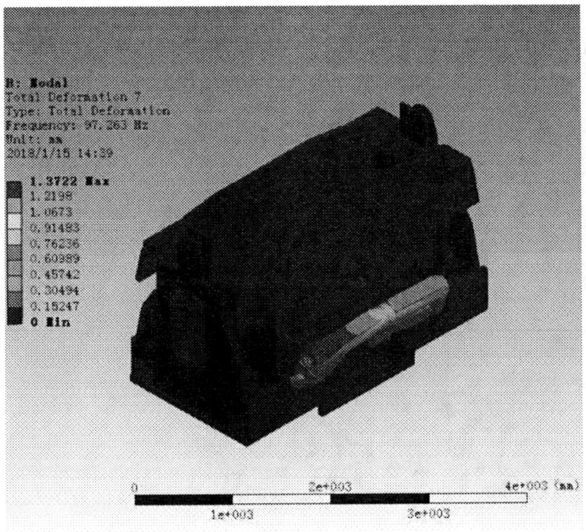

Figure 11. 5th order mode shape cloud chart

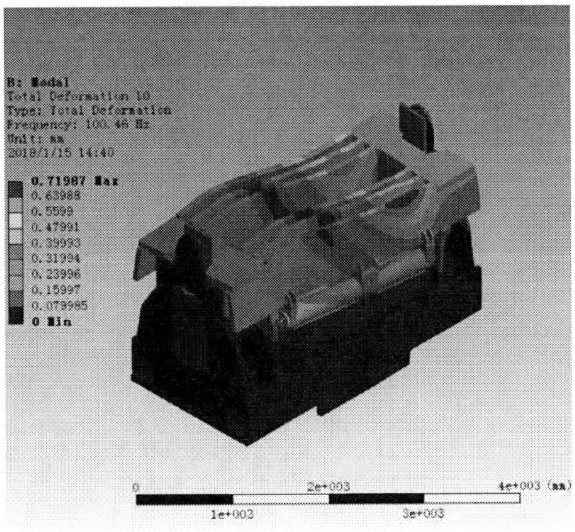

Figure 12. 6th mode shape cloud chart

The dynamic simulation analysis of the segment frame gives the modal vibration modes and vibration frequencies at various stages of the structure [21]. Observing the first six modes of the segment frame, it can be found that the deformation of the whole structure mainly occurs on the nip of the inner arc frame and the outer arc frame. Among them, the deformations in the first, second, third and sixth mode shapes are mainly occurred in the inner arc frame. The deformations in the fourth and fifth modes are mainly occurred in the nip position of the outer arc frame. The maximum deformation of the inner arc frame occurred in the sixth mode modal, the deformation is $0.71987mm$. The maximum deformation of the outer arc frame nip roll occurred in the fifth mode modal, the deformation is $1.3722mm$.

The deformation of the segment frame is directly related to the force of the model and the application of the constraint. Because there is no direct constraint, the inner arc frame has larger swing and torsion. Because of the fixed constraint, the overall deformation of the outer arc frame is very small. However, due to the pressure load exerted on the nip roller, a greater deformation occurs. The frequencies of the first six modes of the segment frame are mainly distributed in the range of $47.384Hz \sim 100.46Hz$. In order to avoid resonance, the influence of the external excitation of this frequency band on the reliability of the equipment should be avoided.

By analyzing the mode shapes of the frame structures at each natural frequency, it can be seen that the dynamic characteristics of the segment frame are good, and the maximum deformation does not exceed the limit value of 1.8mm. The results of dynamic analysis of the segment frame provide the basis for the subsequent optimization of the segment frame of the continuous casting machine.

4 Stiffness test and experimental verification of segment frame

In order to validate the static and dynamic simulation results, experimental equipment was designed to verify the actual maximum deformation of the segment frame.

4.1 Design of stiffness test experimental equipment for segment frame

The hydraulic system has the characteristics of large output force and infinite speed regulation [22]. Therefore, hydraulic jacks were used to simulate the load caused by the static pressure of the slab liquid core on the segment frame, and then the maximum deformation of the frame is measured by the dial indicator. In the experiment, 24 hydraulic jacks were supplied with energy through the hydraulic system. Pressure was applied to the upper and lower driving rollers through the V block to simulate the load of the slab liquid core and the amount of deformation of the segment frame through the dial indicator. Experimental equipment is shown in Figure 13.

Figure 13. Experimental equipment

Hydraulic jack structure is shown in Figure.14. The upper roller and the lower roller are connected with the frame on the segment to play the role of clamping the slab; the hydraulic jack is arranged between the upper roller and the lower roller, and the slab pressure is adjusted and simulated by the expansion and contraction of the piston rod of the hydraulic cylinder; V blocks are placed on both sides of the hydraulic jack, which play the role of connection adapter.

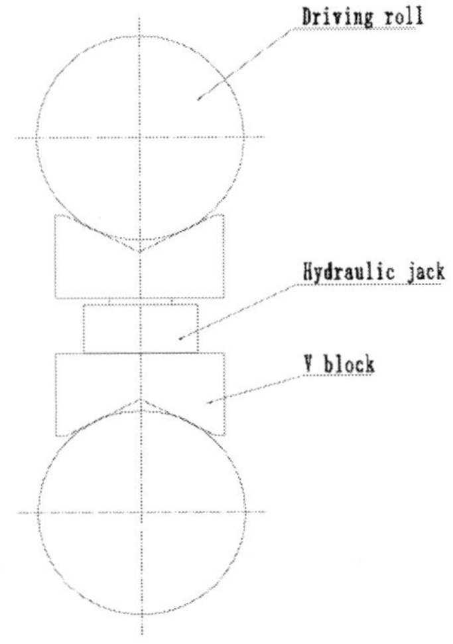

Figure 14. Hydraulic jack structure

4.2 Discussion

First, hang the segment section into the test bench, connect the various parts of the hydraulic pipeline; debug clamping cylinder action, exclude hydraulic cylinder gas, add 300mm standard pad, use system pressure (18*MPa*) to drive. Put the jacks and pedestals into the segment frame and place a jack (except for the driving roller) on each end of each driving roller, and arrange the jack evenly; install the test dial indicator on the side of each jack and place The number of hands adjusted to "0" position; start the experimental hydraulic pump, the pressure from 0 to 1*MPa* to 1.5*MPa*, to eliminate bearing clearance, read the pressure values and record the data; use the 70% / 100% of experimental pressure to withstand the roll surface, so as to ensure good contact between the contact surfaces, record the degree of dial indicator. Finally, remove the test equipment and restore the segment status.

Through many experiments, measure the data of the difference of multiple sets of roll gap through the dial indicator, as shown in Figure15.

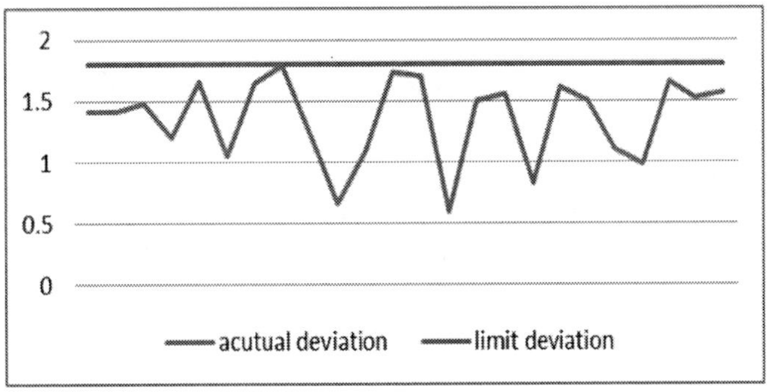

Figure 15. Roll gap actual deviation and limit deviation

The test results show that the maximum deformation of the fan-shaped segment frame occurs at the connection portion of the inner-arc frame pinch roller, where the pinch roller support is weak and the deformation is large. The deformation from the middle of the inner-arc frame to the supporting portion is greatly changed, and the maximum deformation of the roller gap is 1.79*mm*, the minimum deformation is 0.66*mm*, and the results are in good agreement with the simulation and theoretical results.

4.3 Industrial application

After the finite element simulation analysis and experimental verification, with dynamic soft reduction, the centerline segregation has been improved, and the maximum deformation of the segment frame can meet the stiffness requirement. Dynamic pressure implementation of effective compensation, the dynamic configuration of the soft reduction of a variety of different steel slab internal quality improvement effect is obvious. Slab internal quality has been significantly improved, slab segregation less than or equal to Class C 1.5 and eliminate the type A sulfur seal. Obviously, with soft reduction, the centerline segregation of steel slab has been improved, as shown in Figure.16.

without soft reduction with soft reduction

Figure 16. Application effect

5 Conclusion

In order to verify whether soft reduction can improve the stiffness and service life continuous casting machine. In this paper, a three-dimensional model of the segment frame of the continuous casting machine was established. The finite element simulation analysis was carried out to complete the simulation of the maximum deformation and the modal simulation. The simulation results were verified through the hydraulic jack simulation loading test. The experimental results are consistent with the simulation results. The stiffness of the frame structure can meet the requirements with soft reduction.

A three-dimensional solid model of the segment frame was established, and its simplified processing was made to facilitate the finite element simulation. The hexagonal mesh was used to divide the structure so that the calculation accuracy is better and the calculation time is shorter. The maximum deformation cloud, modal vibration frequency and modal vibration modes of the segment frame are obtained through the finite element simulation analysis. On the basis of simulation, the segment frames were loaded through hydraulic jacks, and several experiments were carried out. The experimental results are in good agreement with the simulation results. Segment frame stiffness and dynamic properties are good.

Acknowledgment

The authors gratefully acknowledge the financial support of Grant-in-aid for scientific research from the National Natural Science Foundation of China (No. 51575429).

References

1. K.Ayata, H.Mori, K.Taniguchi, H.Matsuda. Low Superheat Teeming with Electromagnetic Stirring. J, ISIJ Int. **35**, 6 (1995)
2. M.Javurek, M.Barna, P.Gittler, K.Rockenschaub, M.Lechner. Flow Modelling in Continuous Casting of Round Bloom Strands with Electromagnetic Stirring. J, Steel Res. Int., **79**(2008).
3. H.Sun, L.Li, X.Cheng, W.Qiu, Z.Liu, L.Zeng. Reduction in macrosegregation on 380mm×490mm bloom caster equipped combination M+F-EMS by optimizing casting speed. J, Ironmaking Steelmaking, ,**42**(2015).
4. J. C. Li, B. F. Wang, Y. L. Ma and J. Z. Cui. Effect of complex electromagnetic stirring on inner quality of high carbon steel bloom. J, Mater. Sci. Eng. A, **A425** (2006).
5. V. Ludlow, A. Normanton, A. Anderson, M. Thiele, J. Ciriza, J. Laraudogoitia and W. Knoop. Strategy to minimize central segregation in high carbon steel grades during billet casting. J, Ironmaking Steelmaking, **32** (2005).
6. P. Sivesson, T. Örtlund and B. Widell. Improvement of inner quality in continuously cast billets through thermal soft reduction and use of multivariate analysis of saved process variables. J, Ironmaking Steelmaking, **23**(1996).
7. P. Sivesson, G. Häilen and B. Widell: "Improvement of inner quality of continuously cast billets using electromagnetic stirring and thermal soft reduction", Ironmaking Steelmaking, **25**(1998).
8. C. M. Raihle, P. Sivesson, M. Tukiainen and H. Fredriksson. Improving inner quality in continuously cast billets: comparison between mould electromagnetic stirring and thermal soft reduction. J, Ironmaking Steelmaking, 21(1994).
9. P. Sivesson, C. M. Raihle and J. Konttinen. Thermal soft reduction in continuously cast slabs. J, Mater. Sci. Eng. A, **A173 (**1993).
10. W. Bleck, W. Wang and R. Bulte. Influence of soft reduction on internal quality of high carbon steel billets. J, Steel Res. Int., **77(**2006).
11. C. H. Yim, J. K. Park, B. D. You and S. M. Yang. The effect of soft reduction on center segregation in c. c. slab. J, ISIJ Int., **36**(1996).
12. R. Thome and K. Harste. Study on the design of the soft reduction unit of high-speed billet casters. J, Steel Res. Int., **75**(2004).
13. R. Thom and K. Harste. Principles of billet soft-reduction and consequences for continuous casting. J, ISIJ Int., **46**(2006).
14. S. Ogibayshi, M. Kobayshi, M. Yamada and T. Mukai. Influence of soft redution with one-piece rolls on center segregation in continuously cast slabs. J, ISIJ Int., **31**(1991).
15. J. Cappel, R. Flender, R. Ho¨ ffken, G. Kemper and K. Wünnenberg. Centre segregation, soft reduction and oxide cleanness for large diameter line pipe with highest demands on HIC. J, Steel Res. Int., **76**(2005).
16. K. Isobe, H. Maede, K. Syukuri, S. Satou, T. Horie, M. Nikaidou and I. Suzuki. Development of soft reduction technology using crown rolls for improvement of centerline segregation of continuously cast bloom. J, Tetsu-to-Hagane, **80**(1994).
17. L. Morsut. Technological packages for the effective control of slab casting. J, MPT Int., **25**(2003).
18. S. Nabeshima, H. Nakato, T. Fujii, T. Fujimura, K. Kushida and H. Mizota. Control of centerline segregation in continuously cast bloms by continuous forging process. J, ISIJ Int., **35**(1995).
19. Gou Wenxuan. *Mechanics of Materials* (Beijing: Science Press, 2017).
20. Cheng Daxian. *Machinery Design Manual Fifth Edition* (Beijing: Chemical Industry Press, 2008).
21. Wen DeSheng, Lv Shijun, Wen Jia. *New hydraulic system transmission* (Beijing: Chemical Industry Press, 2017).
22. Wenjun Wang, Linxin Ning, Raimund Bülte, Wolfgang Bleck. Formation of internal cracks in steel billets during soft reduction. J, Journal of University of Science and Technology, **15** (2008)

Characterisation of Cellulose Nanofibres Derived from Chemical and Mechanical Treatments

Martini Muhamad[1,a], Peter Hornsby[2], Eugene Carmichael[3], Muhammad Zakaria[4], Yew Been Seok[1], Saiful Mohamed[1] and Shekhar Sharma[3]

[1]*Faculty of Innovative Design and Technology, Universiti Sultan Zainal Abidin, Kuala Terengganu, Malaysia*
[2]*Polymer Cluster, Queen's University Belfast, School of Mechanical and Aerospace Engineering, Ashby Building, Belfast, BT9 5 AH UK*
[3]*Applied Plant Science Division, Agri-Food and Bioscience Institute, Newforge Lane, Belfast, BT9 5PX, UK.*
[4]*Faculty of Manufacturing Engineering Technology, TATI University College, 24000 Kemaman, Terengganu, Malaysia.*

Abstract. Optimised routes have been established to obtain high aspect ratio cellulose microfibrils derived from plant feedstocks, involving a combination of a chemical and mechanical treatment using a high pressure microfluidizer. By this means, nano-cellulose was produced using different pretreatment protocols. The nanofibres produced were characterised for its particle size distribution measurement (laser diffraction and zeta potential analysis) and fourier transform infrared for its structural analysis. Results obtained and subsequent trends observed in chemical, mechanical and chemo-mechanical treated nanofibres were compared and contrasted. The chemomechanical treatment (TEMPO-oxidation and high pressure microfluidisation) yielded higher aspect ratio nanofibrils than nanofibrils made by solely chemical or mechanical treatment.

1 Introduction

There is increasing interest in the use of natural fibres as reinforcing additives for polymers, for example in medical and packaging applications. Nano-variants of these materials can have mechanical properties higher than conventional glass fibres, and potentially of the same order of carbon fibres. Furthermore, since they are derived from renewable sources this makes them more environmentally attractive. These natural fibres have relatively low cost with low density giving high specific mechanical properties to the composite. This type of material has attracted great interest from industry and academia due to the remarkable improvements in properties achievable at very low addition levels compared to virgin polymer or conventional micro- and macro-composites. Although the precise reasons for this improvement have been the subject of intense research, the smaller size of the reinforcing phase undoubtedly increases the probability of structural perfection with less surface defects thereby contributing to a more efficient reinforcement compared to micron-sized alternatives reinforcements [1].

A major drawback of using natural fibres (macro-fibres) for composite applications is resulting in large variation in properties inherent to natural products. The properties could be impacted by climatic

[a] Corresponding author : martinimuhamad@unisza.edu.my

conditions, maturity, soil type and also disturbance during plant growth. Therefore, the idea of decreasing the size of the natural fibre particles from micro to nanoscale has been applied to overcome this problem. The major benefit of using cellulose nanofibres in reinforcing polymer matrices is its high surface area which leads to a very large matrix-filler interfacial area, changing the molecular mobility, relaxation behavior and ensuring thermal and mechanical properties.

The cellulose nanofibres used in this study were derived from banana fibres. For many years, the trunk of the banana tree has been considered as a waste-product and as a result, research was started to evaluate the potential of utilizing the trunk for industrial purposes, in particular after extraction of fibres from the banana trunk. Banana fibre is a ligno-cellulosic fibre, obtained from the pseudo-stem of banana plant (Musa sepientum), which is classed as a bast fibre with relatively good mechanical properties [2]. In preliminary work, the average tensile stress for single banana fibre was recorded approximately 670 MPa.

In this study, work is presented and focused on the optimisation of the processing and derivation techniques to achieve the optimum yield of stable nanofibre suspensions by applying various methodologies: chemical, mechanical and chemo-mechanical treatments. Particular emphasis is given to the structural characterisation of these fibres, in order to define their suitability as potential reinforcements for polymers.

2 Experimental

2.1 Materials

Raw banana fibre obtained from the superposed layers of banana trees pseudostem were produced from COPLACA (Grupo Regional de Cooperativas Plataneros Del Archipielago) and then were processed into a long fine fibres at University of Las Palmas, Canary Island. The long fibres then were cut into short fibres at Grupo Antolin (Grupo Antolin Ingenieria S. A.) and then dried under shade. Average fineness of banana fibre is 2386Nm (4.1 dtex).

2.2 Preparation of cellulose nanofibres

The pre-processed banana fibres were dry-milled and pre-treated with sodium hydroxide, NaOH (1 mol) for four hours at 80 °C. An important aim of this pre-treatment was to increase the surface area of the lignocellulosic material to make polysaccharides more susceptible to hydrolysis [3]. The fibres were then washed and freeze dried before up taken for further three different treatment methods.

2.2.1 Acid hydrolysis

Banana fibres were hydrolysed at different conditions. Pre-treated fibres were added to acid solution which contains 60 wt% of water, 10 wt% of concentrated hydrochloric acid (HCl) and 30 wt% of concentrated sulphuric acid (H_2SO_4). The mixture was heated to 80 °C for 16 hours with continuous stirring and then neutralized using NaOH. The mixture was dialysed against running water for two days.

2.2.2 Mechanical Treatment

The cellulose nanofibres were prepared from banana fibres using a homogenisation process. The pre-treated fibres were first added to de-ionized water and turraxed for 20 minutes at 10 000 rpm. This procedure was carried out by T25 Digital ULTRA-TURRAX® in an ice bath in order to break up the particle size of banana fibres before going through in the microfluidizer. The microfluidiser M-110P is a laboratory unit machine that provides continuous high shear rate processing under high process pressure up to 2868 bar (30 000 psi), with a flow rate up to 120 ml/min and sample size of 50 ml.

2.2.3 Chemo-mechanical treatment

Cellulose nanofibres were prepared using a combination of TEMPO oxidation and microfluidizer. The pretreated fibres were added to an aqueous solution containing 0.000125 wt% of TEMPO (2, 2, 6, 6-Tetramethyl1-piperidinyloxy, free radical) and 0.00125 wt% sodium bromide. The reaction was initiated by adding 0.01 wt% sodium hypochlorite. All the procedures were held at room temperature. The neutral mixture was then passed through into microfluidizer for 10 passes through 200 and 87 μm chamber at 30 000 psi.

2.3 Characterisation of cellulose nanofibres

2.3.1 Dynamic light scattering (DLS)

Nanofibre suspension's particle size distribution was measured by DLS method with a Zatasizer ZS series from Malvern Instruments, England. Nano ZS measures the scattering information close to 180° where the detector position is located at 173°. This is known as backscatter detection where the incident beam does not have to travel through the entire sample.

2.3.2 Zeta potential analysis

Cellulose nanofibres were analyzed by using a Zetasizer ZS series from Malvern Instruments, England. If the particles in suspension have a large negative or positive zeta potential then they will tend to repel each other and no tendency to flocculate. The general dividing line between stable and unstable suspension is generally taken either =30 mV or -30 mV.

2.3.3 FTIR analysis

FTIR spectroscopy is able to provide evidence of any new functional groups, also investigate the chemical structure changes when treatments (chemical and mechanical) are applied. FTIR analysis was performed onto raw materials and cellulose nanofibres in order to study their changes in chemical structure by using attenuated total reflection (ATR) analysis. The FTIR spectra of all specimens were performed using a Perkin Elmer spectrometer (Spectrum 1000) instrument with a universal ATR sampling accessory. 64 scans were collected between 650 and 4000 cm^{-1} with 2 cm^{-1} resolution.

3 Results and discussion

3.1 Size distribution of cellulose nanofibres

Results from DLS determinations are shown in Figure 1, in terms of Z-average diameter. The cellulose nanofibres treated with acid hydrolysis shows a larger particle size compared to the other products, by means that the nanofibres particles treated with acid hydrolysis aggregated in aqueous suspension rapidly compared to other samples. These observations can be explained by higher hydrophobic characteristics of the particles i.e. lower number of OH groups in these samples, leading to aggregation [4]. The average diameter size of cellulose nanofibres particles were 168.9 nm (acid hydrolysis), 85.4 nm (Ultra-Turrax and microfluidizer) and 64.6 nm (TEMPO-oxidation and microfluidizer).

Banana nanofibres treated with a combination of chemical and mechanical treatment shows the lowest average particle size diameter: 64.6 nm. The fibrils obtained from the combination of TEMPO-oxidation and high pressure microfluidisation were much finer compared to the fibrils treated solely with chemical or mechanical treatments. TEMPO-oxidation helps in separating the packed fibrils facilitating subsequent breakdown due to the high shear experienced from microfluidisation. Furthermore, the fibres swollen due to mercerisation treatment should cause higher friction and shear forces in the capillary during microfluidisation, resulting in finer and shorter nanofibrils [5].

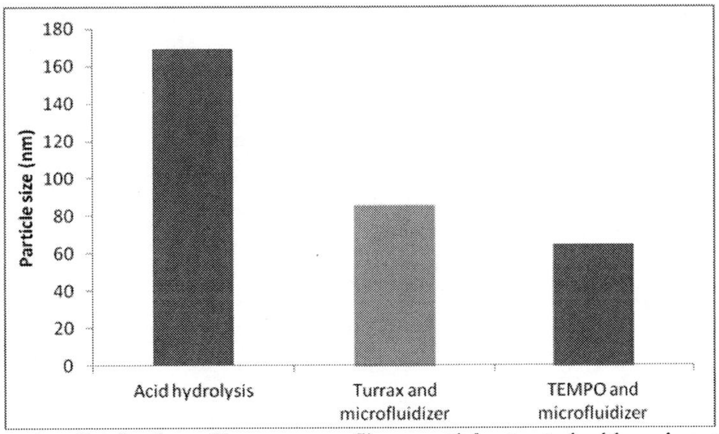

Figure 1. The diameter size distribution of banana nanofibres particles treated with various treatments.

3.2 Zeta potential analysis

Figure 2 shows the negatively charged zeta potential of cellulose nanofibres derived from banana fibre. Cellulose nanofibres prepared with combination of TEMPO-oxidisation and microfluidizer treatment gives the greatest negative zeta potential values: -27 mV showing that these nanofibres were in a more stable dispersed state compared to other samples. This is probably a consequence of the smaller average diameter size particles exhibited from this sample which results in a decrease in particle mass, leading to higher velocity movement and corresponding larger zeta potential.

A recent report by Satyamurthy et al. showed a high zeta potential of -69.7 mV on the surface of cellulose nanowhiskers prepared by sulphuric acid hydrolysis indicating the presence of sulphate groups in the crystallite [6]. This results in stability of the colloidal suspension of cellulose [7-8]. However in the present investigation, the zeta-potential values exhibited from the sulphuric and hydrochloric acid hydrolysed samples were less negative compared to the previously mentioned study. This suggests a lower level of sulphate groups on cellulose nanofibres, a conclusion supported by lack of bound sulphate groups seen in the FTIR spectra at 1210 cm^{-1} that will be presented later.

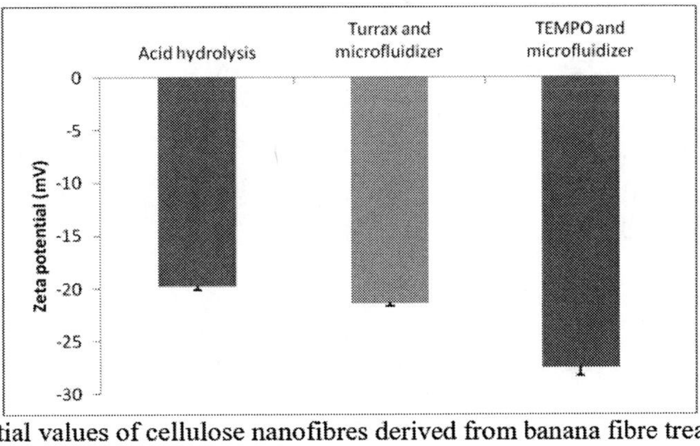

Figure 2. The zeta potential values of cellulose nanofibres derived from banana fibre treated with various treatments.

3.3 FTIR analysis

Figure 3 shows the diffuse reflection FTIR spectra and results for untreated banana fibre and banana nanofibres over 4000 to 650 cm^{-1} wavenumbers. With the nanofibres, there is a significant decrease in peak intensity at 3340 and 2900 cm^{-1}, corresponding to vibration of OH...O hydrogen bonding and antisymmetric and symmetric vibration of CH_2 groups respectively [9]. The presence of peaks at 1445, 1370, 1160 to 1030 and 660 cm^{-1} corresponds to the in-plane deformation of CH_2, in-plane deformation of OH, C-O stretching from ether groups, rocking vibration of CH_2 and out of plane deformation of OH...O hydrogen bonding respectively [10-11]. The presence of hemicellulose and lignin are also evident in the region of 1060 to 1732 cm^{-1} respectively for banana fibre. The disappearance of the small sharp band at 1732 cm^{-1} and general decrease in peak intensities in the region 1060 to 1733 cm^{-1} was observed in banana nanofibres. This indicates that most of the hemicellulose was removed during the chemical and mechanical treatments. The peaks at 1733 and 1242 cm^{-1} in untreated fibre can be attributed to acetyl and uronic ester (C=O) groups from pectin, hemicelluloses along with the ester linkage of carboxylic group of ferulic and ρ-coumaric acids of lignin [9, 11, 12].

However the acid hydrolysis treated nanofibre showed no absorbance in this band and a general decrease in peak intensities in the region of 1600 to 1750 cm^{-1} . This indicates that most of the hemicelluloses and lignin have been extracted during hydrolysis process [4]. The introduction of sulphate groups is presented at 1210 cm^{-1}, however the acid hydrolysis treated nanofibres in this study did not show absorptions corresponding to sulphate groups at 1210 cm^{-1} resulting from neutralisation by NaOH [4].

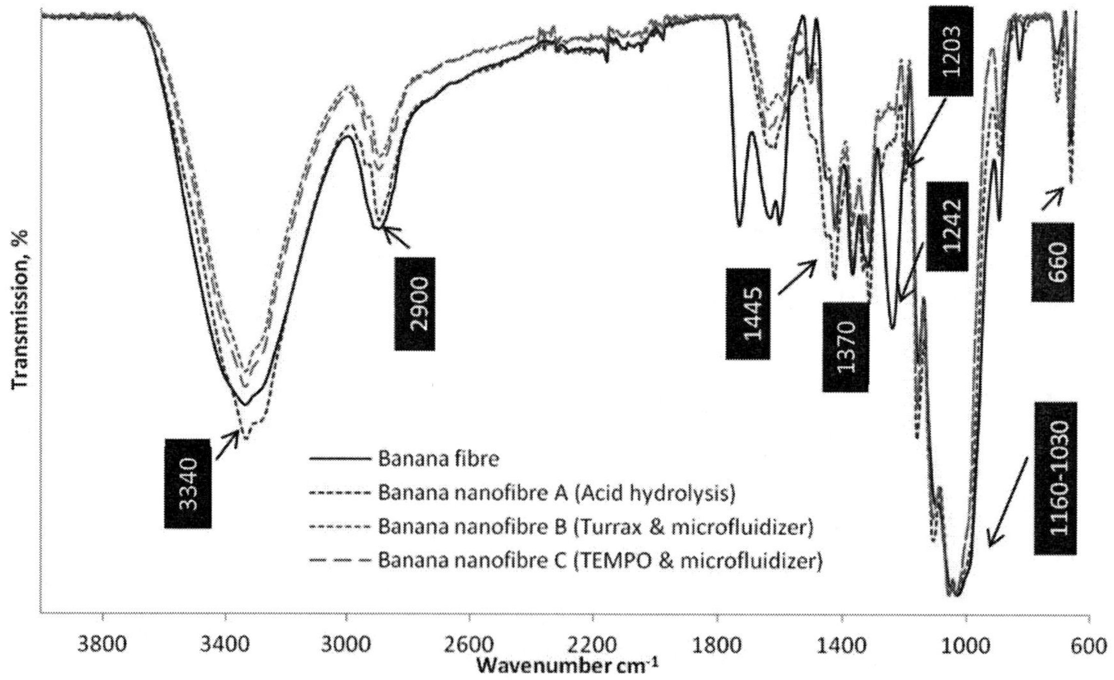

Figure 3. IR spectra of banana fibre and banana nanofibre made by using acid hydrolysis, Turrax and microfluidizer and TEMPO and microfluidizer.

4 Conclusions

The objective of this investigation was to determine an optimum methodology for biorefining of cellulose nanofibres. To this end, cellulose nanofibres were successfully derived from banana fibre using three different strategies; chemical treatment (acid hydrolysis), mechanical shearing (high dispersion Turrax mixer and high pressure microfluidizer) and combination of chemical and mechanical treatments (TEMPO-oxidation and high pressure microfluidizer). Combination of high pressure microfluidisation with TEMPO-oxidation treatment gave the most promising results, resulting in finer cellulose nanofibrils. TEMPO-oxidation was considered to facilitate the fibre breakdown process by reducing the strong intermolecular interaction within microfibril bundles.

Acknowledgment

We wish to acknowledge Research Management, Innovation & Commercialization Centre (RMIC), UniSZA, for the financial support of this work through Special Research Grant Scheme (UniSZA/2017/SRGS/11).

References

1. C. Lew, W. Murphy and G. McNally, "Preparation and properties of polyolefin-clay nanocomposites," *Polymer Engineering & Science,* vol. 44, *(6),* pp. 1027-1035, 2004.
2. S. Mukhopadhyay *et al*, "Banana Fibres - Variability and Fracture Behaviour," *Journal of Engineered Fibres and Fabrics,* vol. **3**, *(2),* pp. 39, (2008).
3. A. Bhatnagar and M. Sain, "Processing of Cellulose Nanofiber-reinforced Composites," *Journal of Reinforced Plastics and Composites,* vol. **24**, *(12),* pp. 1259-1268, August 01, (2005).
4. S. Sharma, "Biorefining of perennial ryegrass for the production of nanofibrillated cellulose," *RSC Adv.,* (2012).
5. A. K. Mohanty, M. Misra and L. T. Drzal, *Natural Fibers, Biopolymers, and Biocomposites.* CRC, (2005).
6. P. Satyamurthy *et al*, "Preparation and characterization of cellulose nanowhiskers from cotton fibres by controlled microbial hydrolysis," *Carbohydr. Polym.,* vol. **83**, *(1),* pp. 122-129, (2011).
7. D. Bondeson, A. Mathew and K. Oksman, "Optimization of the isolation of nanocrystals from microcrystalline cellulose by acid hydrolysis," *Cellulose,* vol. **13**, *(2),* pp. 171-180, (2006).
8. R. Marchessault, F. Morehead and M. J. Koch, "Some hydrodynamic properties of neutral suspensions of cellulose crystallites as related to size and shape," *J. Colloid Sci.,* vol. **16**, *(4),* pp. 327-344, (1961).
9. A. Alemdar and M. Sain, "Isolation and characterization of nanofibers from agricultural residues–Wheat straw and soy hulls," *Bioresour. Technol.,* vol. **99**, *(6),* pp. 1664-1671, (2008).
10. E. Qua *et al*, "Preparation and characterisation of cellulose nanofibres," *J. Mater. Sci.,* vol. **46**, *(18),* pp. 6029-6045, (2011).
11. X. Sun *et al*, "Characteristics of degraded cellulose obtained from steam-exploded wheat straw," *Carbohydr. Res.,* vol. **340**, *(1),* pp. 97-106, (2005).
12. M. Sain and S. Panthapulakkal, "Bioprocess preparation of wheat straw fibers and their characterization," *Industrial Crops and Products,* vol. **23**, *(1),* pp. 1-8, (2006).

Study of Strain-Softening Stage in Structurally Inhomogeneous Materials

Vladimir Mironov[1,2] and Olga Lukashuk[2,a]

[1]*Institute of Engineering Science, RAS (Ural Branch), 620049, 34 Komsomolskaya st., Yekaterinburg, Russia*
[2]*Ural Federal University, 620002, 19 Mira Street, Yekaterinburg, Russia*

Abstract. The paper considers – in terms of modeling and experimental testing – the softening stage manifested in structurally inhomogeneous materials under quasi-static and cyclic load. The method which is used to build the theoretical basis is typical for mechanics. It starts with defining a new material property – strain-softening phase. The stage is realized in the form of a descending branch of the curve on computer diagrams recorded during simple tension or torsion tests on non-reusable specimens in a reasonably rigid testing machine. The type of such a diagram is determined by the structural inhomogeneity of a material, which is defined by distribution law for strength and plasticity properties of structural elements. Complete stress-strain curves or diagrams (CSSD) with a branch descending to zero have been plotted for materials of various classes. Degradation of parameters of a descending branch on a computer diagram recorded for trained specimens provides a new angle on the subject of relation between static and cyclic material properties. The experiments and modeling which were carried out provide basic understanding of softening in terms of deformation.

1 Introduction

The whole variety of material properties and states used in strength calculations on solid bodies is reflected in correlations which define boundary-value problems. As the research on mediums with complex properties progressed, more and more models appeared. Those models allowed to factor in various aspects of body behavior under external load. It was how the theories of elasticity, plasticity, prolonged strength, and many others were formed. According to the internal logic of mechanics, when you try to solve a problem concerning body failure, a new material property should also be defined – strain-softening stage. It is even more important to do when one considers the fact that many attempts to solve the fracture problem by means of elastic-plastic analysis have lead to incorrect results, coming into a conflict with initial experiments [1,2]. Determining such functions of a material which could describe its softening stage is quite promising for the purposes of bridging a known gap between problems and criteria of fracture mechanics and disseminated damage mechanics [2,3] and developing, on that basis, new resource-saving techniques for calculating strengths of machine parts.

2 Modeling of material softening stage

[a] Corresponding author : oldim96@mail.ru

Using even the simplest structural models of Daniels and Masing suffices to single out the stage of material softening from the whole lot [4, 5]. Virtual "dead" tension applied to a bundle of variable-strength parallel rods reveals a branch descending to zero on diagrams plotting relations between ominal stress and strain (stress-strain diagrams). The type of such a diagram is determined by the structural inhomogeneity of a material, which is defined by distribution law for strength and plasticity properties of structural elements. It should be stressed that the more inhomogeneous the material is used, the flatter the descending branch becomes. The importance of this fact can be fully understood only when the behavior of the machine-specimen system is studied (Fig. 1a) using the method of catastrophe theory [6]. Substituting a uniform distribution law for ultimate strengths of elements with a normal law, one is left with a "crease" catastrophe (Fig. 1).

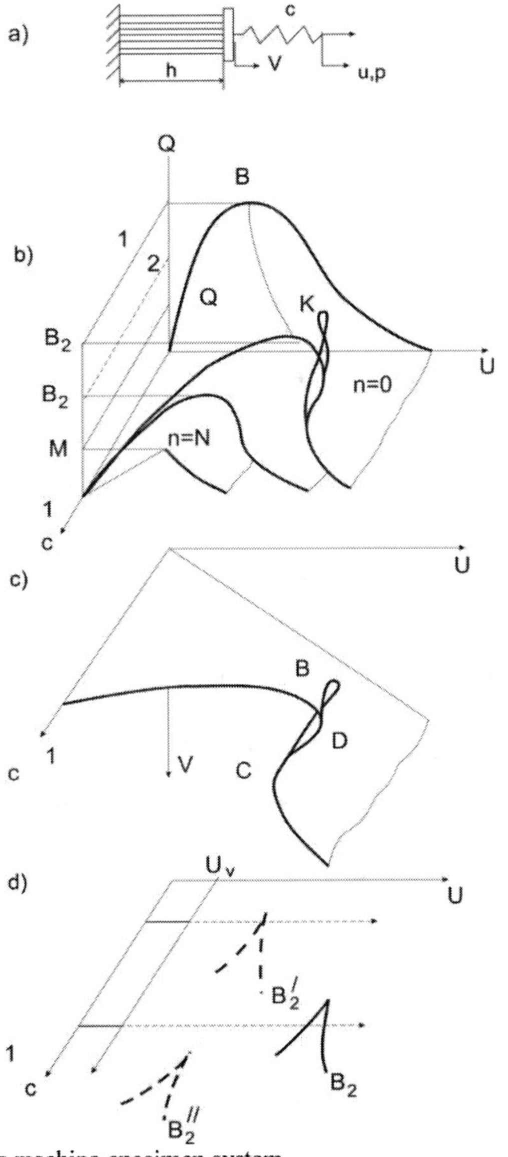

Figure 1. Fold catastrophe of the machine-specimen system.

Analyzing how the surface of equilibrium states of the machine-specimen system behaves in the neighborhood of a bifurcation point (or a set of points) provides a complete scenario of deformation (the cases of diminishing stresses included). Specific surfaces in the form of a crease or fold bear the names of related catastrophes. Figures 1b (for load Q) and 1c (for a preset displacement u of the machine's movable grip) show crease surfaces which are typical of a computer diagram $Q(u)$ with a bifurcation point located on the descending branch. Mapping those catastrophes in the control space reveals areas of stable and unstable behavior of the system (divided by bifurcation lines $B2$ in Fig. 1d).

The instability manifested in the system's deformation in the form of a catastrophic jump on the surface of equilibrium states is related to a complete or partial failure of a specimen. For example, when a specimen is loaded with a certain force, it leads to its dynamic failure at the maximum of the computer diagram regardless of the rigidity of the loading device (line *B-K* in Fig. 1b). To the contrary, when kinematic loading is used, the machine's rigidity plays a crucial role in providing stability of the system's deformation. At a sufficiently high rigidity, there are ways of achieving stable deformation, or even equilibrium (without dynamics) division of a specimen into two parts (path 1 in Fig. 1d). When the rigidity of the loading system is lowered, bifurcation points come up in the strain path, which leads to a jump from one plane of the surface Π to another (path 2 in Fig. 1d). Theoretically, in the case of an infinitely small rigidity, the system would lose its stability at the maximum of the computer diagram – similar to the above case of loading with a specified force. Obviously, the fracture pattern for specimens made of the same material would vary depending on the rigidity of the loading system – dynamic (or brittle) when the rigidity is low, and equilibrium (or ductile) when it is high.

The deformation instability of the system regarded as a criterion of a specimen's fracture also takes into consideration the impact which the material's properties have on the fracture pattern. It shows up in the dependence of the system's stability conditions on the material's structural inhomogeneity. As model-based calculations demonstrate, the structural inhomogeneity determines the slope of the descending branch, which, in turn, determines the location of bifurcation lines B_2 in the plane of control parameters (Fig. 1d). The more homogeneous the material is, the more abrupt the descending branch of the relation diagram is, and the higher the rigidity of the loading system should be to ensure an equilibrium fracture of a specimen. Thus, to correctly solve problems of fracture mechanics, instead of a material constant, a material function describing the softening stage is needed, along with a methodology which would take the conditions of loading into consideration.

A material function in the form of a diagram which relates stress and strain (deformation) to the descending-to-zero branch is called a function of resistance or a complete stress-strain diagram (CSSD) [6]. One important practical area of usage for such functions is connected to description of prolonged fracture. When parameters of a static resistance function change due to cyclic material damage, it leads to changes in bifurcation pattern of a studied machine-sample system. Depending on controlled parameters of the cycle and behavior of the CSSD degeneracy, there are several scenarios of fatigue failure. The catastrophe could draw closer to the path of loading ($B_2^{/}$ in Fig. 1d) or move away from it ($B_2^{//}$ in Fig. 1d). In spite of those two different scenarios of fatigue process, the final failure of a specimen happens when the machine-specimen system loses its stability. Therefore, the instability of strain could be defined as a unified cause of static and cyclic failure of specimens. Only conditions of its occurrence are different.

Modeling turned out to be quite effective in terms of finding patterns of change in material properties at the stage of softening. Structural material models were additionally supplemented with conditions of component failure. It was assumed that an element which has been fractured at the point of its ultimate strength moves apart, without pushing, those elements which have not been affected, thus creating a void. When load is withdrawn, this void could close up, with the fracture element working in compression. Such models were used to study instantaneous or tangent material properties under active added loading E^p, ν^p and unloading E^u, ν^u, necessary to form incremental physical equations (E is a modulus, and ν is a lateral deformation coefficient). Fig. 2 shows the results of numerical experiments when the normal law of distribution is maintained for deformations of elements transition into the plastic ε_T and fracture ε_Z states.

For a model specimen as a unified system of N elements,

$$\varepsilon_k = \varepsilon, \sigma = \frac{1}{N}\sum_{k=1}^{N}\sigma_k, \eta = \frac{1}{N}\sum_{k=1}^{N}\eta_k \qquad (1)$$

is true in the case of strain, and

$$\varepsilon_k^u = \varepsilon^u, \sigma^u = \frac{1}{N}\sum_{k=1}^{N}\sigma_k^u, \eta^u = \frac{1}{N}\sum_{k=1}^{N}\eta_k^u \tag{2}$$

- when unloading takes place.

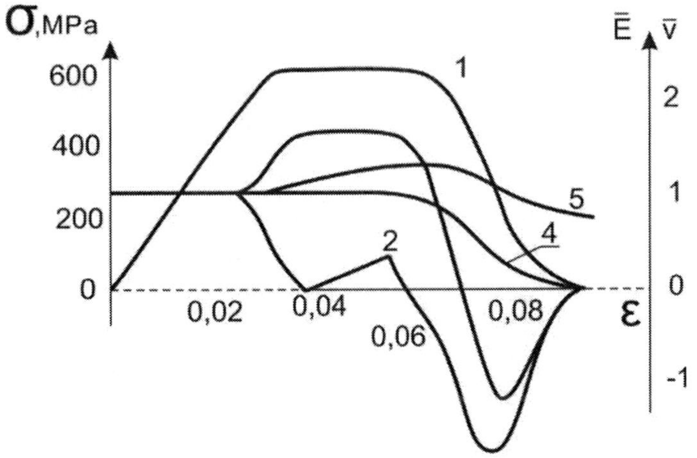

Figure 2. Properties of model material.

These expressions helped to find, first of all, original dependences $\sigma(\varepsilon)$, $\eta(\varepsilon)$, $\sigma^u(\varepsilon^u)$, $\eta^u(\varepsilon^u)$, and then, via numerical differentiation, target functions $E^u(\varepsilon^u) = \frac{d\sigma^u}{d\varepsilon^u}, \nu^u(\varepsilon^u) = -\frac{d\eta^u}{d\varepsilon^u}$ and $E^u(\varepsilon^u) = \frac{d\sigma^u}{d\varepsilon^u}, \nu^u(\varepsilon^u) = -\frac{d\eta^u}{d\varepsilon^u}$, manifested in the form of deformation functions at the start of unloading ($\varepsilon = \varepsilon^*$).

The dependence $\sigma(\varepsilon)$ in the Fig. 2 (curve 1) has a descending branch which comes up as a result of separate elements of the system being fractured and characterizes the strain-softening of the model material. Graphs E^u, ν^u, which are drawn in relative coordinates (line 4), match and change only when structural elements are fractured. To the contrary, current values of E^p, ν^p begin to change when plasticity appears, and by the start of fracture ($\varepsilon = \varepsilon_{z\alpha}$), would have taken values of $E_p = 0$, $\nu_p = 0.5$. One could point out at the similar pattern of change for the graphs E_p and ν_p (lines 2` and 3 in the Fig. 2) when $\varepsilon > \varepsilon_{z\alpha}$. The tangent modulus becomes negative, and the lateral strain coefficient diminishes and falls down to negative values, as well, which reflects the fact of considerable loosening of the material. Line 5 (Fig. 2) corresponds to the lateral strain coefficient, which could be defined as a correlation between complete lateral and axial strains. If residual plastic deformations (typical of this model) are present, the finite value of the coefficient is above zero.

The modified Masing model could be extended by adding cyclic properties of materials – for example, the law of random longevity distribution and the pattern of ultimate strength degeneracy [7]. In that case, when stationary loading is applied to a model specimen, one can witness gradual degeneracy of resistance function, along with every kinetic curve mentioned here.

3 Experimental Study of Strain-Softening Stage

Phenomenological approach to the problem of fracture predetermines the methods of studying material properties based on mechanical testing of specimens. At the same time, the specifics of carrying out the tests aimed at recording CSSDs touch upon fundamental notions of experimental mechanics: hypothesis of continuity and macroscopic definability. The conditions for specimen testing and their geometry contradict rationality of established experimental principles. New demands are made not only on technology and tools of experiments but also on analysis and interpretation of the results obtained.

Strictly speaking, an object to be tested should have a typical or representative volume of material [4]. Departure from this requirement in the field of elastic vibrations is justified by the fact that one can manage to obtain such characteristics of material which would be invariant to the size of the object in question. But even in typical tests – for example, in uniaxial tensile tests – their output characteristics does depend on the size of a specimen [8,9]. Plotting CSSDs aggravates this problem due to inevitable localization of deformations by the length of the specimen. The working section of the specimen which is outside the area of localization becomes a part of the loading system, largely diminishing its rigidity. As a result, one fails to plot the descending branch on the computer diagram when typical specimens are used, even in the case of a high-rigidity loading device.

In accordance with modeling principles, it is necessary to test specimens of smaller sizes, if possible. In uniaxial tensile tests, one-time specimens of 2 mm in diameter and 0,3 mm in radius of fillet transition were utilized as having close-to-standard parameters of the ascending branch. A ring-like loading device of variable rigidity allowed to draw a CSSD with a branch descending to zero for materials of various classes [10,11]. Recalculating the functions of resistance using stress-strain coordinates (based on Koshi) of computer diagrams allowed to plot CSSDs for materials of various classes (Fig. 3) as specified by Russian standards.

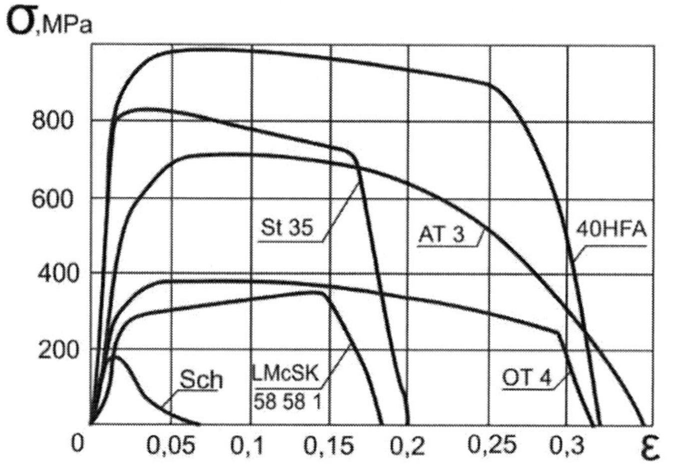

Figure 3. Complete Stress-Strain Diagrams.

Building one model or another of a material in the state of softening predetermines selection of characteristics which would reflect the destructive processes taking place within [12-16]. One could assume free-energy release and load redistribution among elements of the medium to be typical signs of fracture. Two mechanisms, at least, are in charge of energy dissipation in materials: plastic deformation and formation of structural micro-defects. The mechanical characteristics corresponding to those would be the ultimate strain ε_Z (determined by the length of a CSSD) and the modulus of unloading E'' (diminishing as formation and development of defects progress). Synergetic material properties, which are related to redistribution of stresses when defects appear and grow, are determined by structural inhomogeneity. The slope of the descending branch of a CSSD is also connected to the structural inhomogeneity. Besides, it closely correlates with crack growth resistance of a material [13]. The corresponding mechanical characteristic would be a $E^p(\varepsilon)$ function. Another indicator of the degree by which the material is damaged could be a $\nu^P(\varepsilon)$ function. All functional material characteristics noted above are derived from a basic experiment on plotting a CSSD with interim unloading points.

The function of resistance with a descending-to-zero branch gives the most comprehensive and logically complete idea of static material properties. It also provides a unifying concept of the limiting state of a material, relating it to the loss of stability in the process of deformation. In tests run on cast-iron specimens, which would manifest no neck until their fracture, patterns revealed were similar in terms of quality to the ones in Fig. 2 [11].

4 Conclusion

Regarding fracture as a loss of deformation stability leads to conceiving a new notion of the limiting state of a material. The descending branch of the resistance function provides a full set of marginal material states for a specified path of loading. In certain experimental conditions, all testing runs in equilibrium, without a loss of stability of the machine-specimen system. When conditions of stability fail in a specific case, only a part of the descending branch is followed in equilibrium. The state of the material at a critical point when the body loses its deformation stability could be called a critical limiting state.

Experimental studying of the phenomenon of material softening mainly follows major principles of experimental mechanics, but certainly it has its own peculiarities and methodological issues. This statement is true at least for structural materials which have no neck when strained.

Degradation of the parameters of the descending branch of a computer diagram for specimens which have undergone cyclic training provides a new angle on the subject of relation between static and cyclic material properties. In both cases, the loss of deformation stability plays the roles of both a cause and a criterion of fracture. The results cited here give a certain basis for further modeling and experimental studies into the softening stage in structural materials.

References

1. G.P. Cherepanov, *Mechanics of Brittle Fracture* (Science, Moscow, 1974)
2. S.D. Volkov, V.P. Stavrov, *Statistical Mechanics of Composite Materials* (BSU Publishing, Minsk, 1978)
3. J.A. Callins, *Failure of Materials in Mechanical Design: Analysis, Prediction, Prevention*, (World Moscow, 1984)
4. S.D. Volkov, *Statistical Theory of Strength*, (Machine Building State Publishing, Moscow, 1960)
5. S.D. Volkov, *Function of material strength and statement of boundary problems of fracture mechanics*, (USSR Academy of Sciences, Sverdlovsk, 1986)
6. V.I. Mironov, O.A. Lukashuk, J. Key Eng. Mater., **735**, 89-94 (2017)
7. V.I. Mironov, A.V. Yakushev, J. Bull. of USTU, **22** (2004)
8. N.A. Shaposhnikov, *Mechanical Testing of Materials*, (Mashinostroenie, Moscow, 1954)
9. V.T. Troshhenko, L.A. Sosnovskiy, *Fatigue Strength of Metals and Alloys* (Naukova Dumka, Kiev, 1987)
10. A.S.G 01 N 3/08 *Device and method of testing material specimens on strain*, publ. 10.05.2005. bul. 13, 2005.
11. V.I. Mironov, O.A. Lukashuk, D.I. Vichuzhanin, J. Sol. St. Phen., **815**, 815-820 (2017)
12. A.A. Lebedev, N.G. Chausov, J. Strength Problems, **2**, 6-10 (1983)
13. I.G. Emel'yanov, V.I. Mironov, *Durability of Shell Structures* (UrO RAN Publishing, Yekaterinburg, 2012)
14. V.F. Terentiev, *Fatigue Strength of Metals and Alloys* (Intermetals Engineering, Moscow, 2002)
15. G.V. Uzhik, *Fatigue and Endurance of Metals* (Foreign Literature Publishing, Moscow, 1963)
16. V.V. Struzhanov, V.I. Mironov, *Strain-softening of Materials in Structural Elements* (UrO RAN Publishing, Yekaterinburg, 1995)

MATEC Web of Conferences **253**, 01004 (2019) https://doi.org/10.1051/matecconf/201925301004
MSME 2018

Two Approaches to the Problem of High-cycle Fatigue of Materials and Structures

Vladimir Mironov[1,2], Olga Lukashuk[2,a] and Dmitriy Ogorelkov[1,2]

[1]*Institute of Engineering Science, RAS (Ural Branch), 620049, 34 Komsomolskaya st., Yekaterinburg, Russia*
[2]*Ural Federal University, 620002, 19 Mira Street, Yekaterinburg, Russia*

Abstract. Creation of new types of materials and development of new methods aimed at extending the lifetime of structures are two interconnected approaches to solve the problem of metallic structures fatigue. The paper considers a method for estimating the efficiency of damping implemented in metallic structures by means of energy-intensive materials. The results of comparative calculation are given on the ultimate number of transverse vibration cycles in bridge girders. The calculation uses a linear hypothesis of fatigue-caused damage summation and a model of cyclic material degradation. Notable increase of the vibration decrement for foam-filled metallic structures is predicted, with the lifetime of products lengthening as a result.

1 Introduction

The problem of extending the lifetime of metallic structures of lifting and transporting machines is most often solved via technological techniques aimed at affecting the surface layer of metals or by means of diminishing variable dynamic loads. The latter method often uses structural damping [1-4]. The paper [5] discussed one of the approaches to increasing the durability of a thin-walled closed-profile beam by filling it with a low-modulus energy-intensive material. A method of evaluating the lifetime of structural elements which takes cyclic degradation of material properties into account was analyzed in [6,7]. This study uses that method for the purposes of comparatively estimating the lifetimes of unfilled and foam-filled bridge girders of an overhead crane.

The efficiency of structural damping applied to a metallic structure is assessed by running a costly experiment on the basis of a vibrorecord for free damped vibrations. Below a possibility is considered of theoretically predicting the lifetime of a filled bridge girder from a known logarithmic vibration decrement for unfilled beams. For example, the decrement δ_1 for the span of a bridge crane in [8] is evaluated to be in the range of $\delta_1 \in [0.05;0.12]$. Thus, the decrement δ_2 for filled beams could be found using the basic theory of vibrations for a material particle with its mass reduced to the centre of the span.

2 Theoretical estimation of logarithmic decrement δ_2

[a] Corresponding author : oldim96@mail.ru

Differential equation for damped vibrations of a material particle could be expressed as [9]

$$\ddot{x} + 2n\dot{x} + p^2 x = 0 \tag{1}$$

where coefficient n and circular frequency p are linked to a circular frequency p_0, determined for free sustained (undamped) vibrations by the correlation $n^2 = p_0^2 - p^2$.

The decrement δ, conditional period of vibrations T and circular frequency are interconnected via the following expression:

$$\delta = nT = n / p. \tag{2}$$

Denoting the vibration parameters for an unfilled girder by 1 in their lower indices and for a filled girder – by 2, and omitting insignificant effect of a light-weighted filler on the circular frequency p_0, the next two expressions could be formulated:

$$n_1^2 = p_0^2 - p_1^2;\ n_2^2 = p_0^2 - p_2^2.$$

Then, elimination of p_0 gives us the expression $n_2 = \sqrt{n_1^2 + p_1^2 - p_2^2}$, while taking Eq. 1 into account and allowing for the fact that a vibration frequency ω relates to the circular frequency as $p = 2\pi\omega$ helps us to obtain the desired expression in the form of

$$\delta_2 = \frac{1}{\omega_2}\sqrt{(\delta_1\omega_1)^2 + \omega_1^2 - \omega_2^2} \tag{3}$$

The vibration frequencies ω_1 and ω_2 for unfilled and filled beams could easily be determined using a typical computer software, for example, such as SolidWorks [10,11]. When the decrement δ_1 is pre-determined, Eq. 3 is completely defined.

The solution of Eq. 1 for a known coefficient n provides the law of variation in point-mass displacements, and, for a linear system, even the law of variation in damped-vibration stresses

$$\sigma = \sigma_0 \exp(-nt) \tag{4}$$

where σ_0 is a dynamic stress related, for example, to the acceleration applied to or removed from the system when starting to lift the load or applying the brakes. Constraining the level of significant amplitudes to the fatigue limit σ_R, the period of their action t_* could be found on the condition of $\sigma_0 \exp(-nt_*) = \sigma_R$. Then, the number of damped vibrations caused by a single dynamic load which produces the effect of σ_0 stress in a critical point of a metallic structure would be expressed as

$$N_Z = \frac{t_*}{T} = \frac{-\ln(\sigma_R / \sigma_0)}{Tn} = \frac{-\ln(\sigma_R / \sigma_0)}{\delta} \tag{5}$$

In engineering calculations, σ_0 means the normal stress in a critical section of a beam [8,12] or the von Mises stress [13]. The latter is determined in terms of high-cycle fatigue by solving a boundary value problem from the theory of elasticity. The SolidWorks software package provides desired numerical solution using the finite-element method.

3 Model of material accounting for cyclic degradation

For a specified law of variation in alternating loads, the durability could easily be predicted on the basis of a linear hypothesis of fatigue-caused damage summation [14]

$$\sum \frac{n}{N} = 1 \tag{6}$$

But, neither the measure of damage n/N, nor the criterion of failure (equaling 1) do not have any physical explanation. Correcting the right part from experimental data does not guarantee to have the accuracy of prediction maintained when the conditions of testing change. The authors base their approach to the problem of metal fatigue [7] on the experiments aimed at degrading such parameters of stress-strain diagrams as ultimate strength and ultimate strain. With a force approach chosen to describe the fatigue process and an exponential function picked to approximate an experiment-based kinetic curve, this expression follows:

$$S_B(\sigma_M;n) = S_{B0} - k_\sigma n^m \tag{7}$$

where S_{B0} is ultimate strength, σ_M is maximal stress in a reduced cycle, and the expression $k_\sigma n^m$ has been taken in the form of a Corten-Dolan damage function, which has its own experimental verification. Then, k_σ could be determined on the condition of failure

$$S_B(\sigma_M;N_M) = \sigma_M \tag{8}$$

playing the role of the criterion for the fatigue failure in the case of regular cyclic loading with the expression of Eq. 7 taking the form of

$$S_B(\sigma_M;n) = S_{B0} - \frac{S_{B0} - \sigma_M}{N_M^m} n^m, \tag{9}$$

where N_M is the durability by the Weller diagram, m is an experimental material constant.

Generalization of the approach onto nonstationary load is defined on the condition of equivalence of two cyclic material states with a different history of loading:

$$S_B(\sigma_1;n_1) = S_B(\sigma_2;n_2) \tag{10}$$

This equation helps to find an equivalent (in terms of damage) number of cycles $n_э$ in that kinetic curve to which the transition is made. The criterion of fatigue failure in Eq. 8 is generalized onto the nonstationary load in the form of a rule for intersection of an actual material strength and the level of maximal stress in the cycle.

How the lack of uniaxiality in the stressed state affects the lifetime has not been studied in a sufficient detail. It is known, for example, that the second main stress did not impact the lifetimes of pipe specimens if it was less than 0.8 of the main stress itself [15]. In that case, the kinetic curve from Eq. 7 could be used immediately in phenomenological calculation.

Eqs. 7-10 form the original part of the model of material under cyclic loading and fit known expressions of the classic continual theory of fatigue. This fact allows to include into the model experimentally-verified results of long-term studies on the topic of materials and structures fatigue strength. For example, classification of components in terms of stress concentration and a table of basic strength limits σ_{-1KB} factoring in the quality class for welded seams [16] certainly would be useful when designing welded metallic structures.

An important correction into the procedure of evaluating the ultimate strength σ_{-1K} for a *t*-thick element is introduced by the scale effect

$$\sigma_{-1K} = \sigma_{-1KB}\left(\frac{t_B}{t}\right)^{0,2} \tag{11}$$

where t_B is thickness of a basic specimen.

The fatigue curve could be used both in its typical form

$$\sigma_R^\alpha N_B = \sigma_i^\alpha N_i, \tag{12}$$

and in any other form derived from the results of mechanical testing on the material fatigue. On the other hand, the non-linear function of Eq. 7 presumes that the damaging action of the cycle of stresses depends on where they are located in the whole spectrum of loading. Therefore, regrouping of cycles

– which is a common practice in correlation and spectrum analyses of loading – is excluded. To illustrate the point, self-induced transverse vibrations of bridge girders of an overhead crane should be considered. Those emerge at the moments of starting and stopping the travelling mechanism and come in two variants – for unfilled and foam-filled girders. The symmetry in the cycle of loading (with a constant asymmetry coefficient R=-1) is provided by varying stresses in accordance with the law of Eq. 4.

4 Vibrations of bridge girders of overhead cranes

The dimensions of a bridge girder for an overhead crane of 19.5 *m* in span and 20 *t* in lifting capacity were found in typical strength and rigidity calculations. That beam is 1.1 *m* high and 0.6 *m* wide, and the required thickness of its wall is 9 *mm*. A finite-element model of an unfilled beam was built in SolidWorks (Fig. 1), followed by an assembly model with a filler.

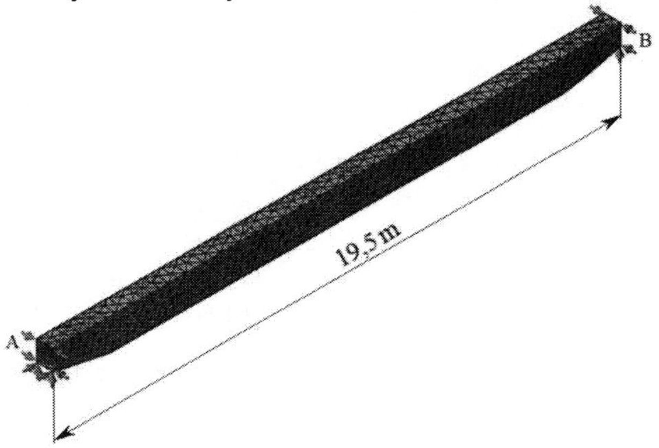

Figure 1. Finite-element model of bridge girder.

Foam was chosen to act as a filler, having the following characteristics: modulus of elasticity – 0.5 *MPa*, Poisson coefficient – 0.49, mass density – 10 kg/m^3. Lower frequencies in the horizontal plane were obtained as $\omega_1 = 7.0546 s^{-1}$; $\omega_2 = 6.980 s^{-1}$. The efficiency of structural damping implemented in a beam could be evaluated on the logarithmic decrement δ_2, which is calculated using Eq. 5 for various values of the logarithmic decrement δ_1 for an unfilled beam (Fig. 2).

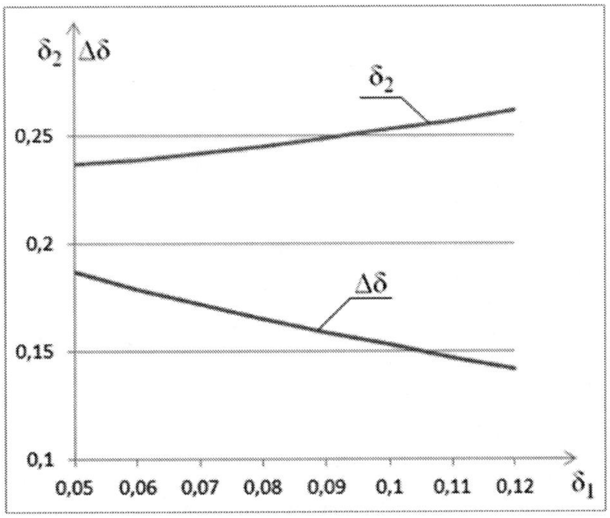

Figure 2. Efficiency of structural damping.

If the efficiency is estimated on the expression $\Delta\delta = \delta_2 - \delta_1$, one should note that it decreases with the increase of δ_1. The graphs above clearly illustrate that damping bridge girders by means of foam materials is certainly quite efficient. The effect also shows up in the number of significant vibration amplitudes $N_{\bar{}}$ in the case of a one-time application of dynamic load. For example, the number dropped from 18 to 6 vibrations for $\delta_1 = 0.05$. It could be demonstrated that the damping-caused decrease of the amplitude and the number of self-induced vibration cycles $N_{\bar{}}$ (determined by Eq. 3) leads to the extension of the product lifetime.

The loads for starting and braking operational modes of the crane were evaluated using a dynamic model of the beam travelling alongside the trolley and load itself. That evaluation provided the following results: maximal normal stresses in the beam when the crane stopped – σ_0=120 *MPa*; constrained fatigue limit for $N_B = 2\cdot10^6$ cycles, welded seams and the scale effect of Eq. 11 – σ_R =50 *MPa*; indicator of the fatigue curve from Eq. 12 for the St3sp3 steel – α =5,34 [17].

When the maximal stresses in the cycle change following the law of Eq. 4, a cycle-by-cycle alteration of stresses occurs. One block of such stresses consists of $N_{\bar{}}$ damped-amplitude cycles. And for the *i*-th cycle, the following expression is true:

$$\sigma_i = \sigma_0 e^{-n\mathrm{T}i} = \sigma_0 e^{-\delta i} \tag{13}$$

When Eqs. 8 and 9 are considered together, it leads to the expression which allows to calculate an actual value of the ultimate strength after a certain equivalent number of cycles

$$S(\sigma_{M1}, n_E + \Delta n_1) = S_{B0} - \frac{S_{B0} - \sigma_{M1}}{N_{M1}^m}(n_E + \Delta n_1)^m, \tag{14}$$

where $\Delta n = 1$ for the cycle-by-cycle variation in stress amplitudes (as in the example).

Based on this equation, an algorithm was developed for comparative calculation of the number of cycles prior to the failure of a metallic structure due to transverse vibrations. The algorithm uses the linear hypothesis of Eq. 6 and the model of cyclic material degradation described. For a number of steels, a numerical value of the constant m was close to 2, and, therefore, the calculation assumed m = 2. The numbers of cycles prior to the failure due to transverse vibrations for the linear damage summation NLH and in terms of the proposed model of material ND (Table 1) are expressed in blocks of vibrations caused by a single dynamic load.

Table 1. Results of computing experiments.

Decrement		N_{LH}	N_D
δ_1	0.05÷0,12	5716÷16514	4302÷8692
δ_2	0.155÷0,19	23510÷31743	10364÷11528

The table shows the interval boundaries of variation in the logarithmic decrements δ_1 and δ_2 (calculated using Eq. 3). The second column provides lifetimes in terms of cycle blocks calculated using the linear hypothesis for the specified boundary values of the decrements in the first column. Those numbers indicate that the structural damping affects notably the lifetimes. The third column gives lifetimes for the same decrements calculated using the model which allows for the cyclic degradation of material. A substantial discrepancy is observed between the predictions of the ultimate number of cycles, and it is especially the case with smaller values of the decrement δ_1 for unfilled structures.

5 Summary

1. The approach is given to estimating the logarithmic decrement for metallic structures filled with an energy-intensive material on the basis of a pre-determined decrement found in literature sources.
2. Large increase of the decrement is predicted when a foam material is used as a light-weighted filler for metallic structures with closed-profile cross-sections. This relatively simple and reliable type of structural damping could prove to be quite competitive in terms of comparison with other appliances meant to damp vibrations (such as electromechanical or hydromechanical systems).
3. Fast damping of vibrations is vital for many designs, including those where precise positioning is important, and, in particular, in the case of bridge girders of an overhead crane. The calculated number of significant self-induced vibrations of such girders decreases by two or three times, which greatly affects the lifetimes of their metallic structures.
4. The prediction of the ultimate number of cycles for transverse vibrations using a model of cyclic material degradation substantially differs from the prediction which uses a linear hypothesis of fatigue damage summation, with shorter lifetimes predicted by the model.

References

1. V.N. Kobelev, L.M. Kovarskiy, S.I. Timofeev, *Design of Three-Layer Structures* (Mashinostroenie, Moscow, 1984)
2. K.V. Frolov (Ed.), *Protection from Vibrations and Impacts* (Mashinostroenie, Moscow, 1981)
3. Singiresu S. Rao, *Mechanical Vibrations* (Addison-Wesley Pung Company, 1995)
4. On-line ANSYS Theory Reference, Release 9.0
5. V.I. Mironov, O.A. Lukashuk, D.A. Ogorelkov, *Proceedings of MATEC Web Conference* (2017)
6. V.I. Mironov, O.A. Lukashuk, D.I. Vichuzhanin, J. Sol. St. Phen., **815**, 815-820 (2017)
7. I.G. Emel'yanov, V.I. Mironov, *Durability of Shell Structures* (UrO RAN Publishing, Yekaterinburg, 2012)
8. S.A. Sokolov, *Structural Mechanics and Metallic Structures of Machines* (Politehnika, Saint-Petersburg, 2011)
9. Y.G. Panovko, *Basics of Applied Theory of Elastic Vibrations* (State Science and Technology Publisher of Engineering Literature, Moscow, 1957)
10. A.A. Alyamovskiy, A.A. Sobachkin, E.V. Odintsova, et al., *Computer-aided Design in SolidWorks for Engineering Purposes* (BHV-Peterburg, Saint Petersburg, 2005)
11. K.A. Basov, *ANSYS. User's Manual* (DMK-Press Publ., Moscow, 2005)
12. *Standard for calculationand designing of cars of railways of Ministry of Railways of a track of 1520 mm (not automotive).*, (GosNIIV-VNIIZhT, Moscow, 1996)
13. Godunov S.K., J. Russian Mathematical Surveys, vol.16, **3**, 171 - 174 (1961)
14. A. Miner, J. Appl. Mech., **12**, 159-164 (1945)
15. N.I. Grinenko, *Proceedings of School-Seminar on Fatigue Strength of Machine-Building Structures* (1975)
16. V.I. Trufyakov, *Fatigue of Welded Joints* (Naukova dumka, Kiev, 1973)
17. V.T. Troshhenko, L.A. Sosnovskiy, *Fatigue Strength of Metals and Alloys* (Naukova Dumka, Kiev, 1987)

MATEC Web of Conferences **253**, 01005 (2019) https://doi.org/10.1051/matecconf/201925301005
MSME 2018

Residual Stress Distribution Analysis in Advanced Materials by Neutron Diffraction: The Case of Spherical Storage Tank Butt Weld

Charles Hervoches[1,a], Miroslav Vrána[1], Pavol Mikula[1], Ľuboš Mráz[2], Peter Baksa[2] and Milan Baláž[2]

[1]Nuclear Physics Institute ASCR, v.v.i. 250 68 Řež, Czech Rep.
[2]Welding Research Institute – Industrial Inst., Račianska 71, 832 59 Bratislava, Slovak Rep.

Abstract. The two-axis diffractometer SPN-100 at the Center of Accelerators and Nuclear Analytical Methods (CANAM) in Řež, is a neutron diffraction instrument dedicated to residual stress analyses in polycrystalline materials, it has recently been equipped with a new two-dimensional position sensitive detector (2D-PSD) and a six-axis robotic arm for sample positioning. Recent results of strain/stress distribution measurements with the upgraded tools are presented. The measured materials are made of C-Mn unalloyed steel used for spherical storage tank. Welding-induced residual stresses have direct influence on ductility and toughness of both weld metal and heat affected zone of the weld area, which affect the service life time of structures. Normal, transversal and longitudinal components of the stresses obtained in the vicinity of the butt weld from a spherical storage tank were measured by neutron diffraction at CANAM.

1 Introduction

Residual stresses are phenomena that develop in all welded structures due to local heating. Stresses have direct influence on ductility and toughness of both weld metal and heat affected zone of the weld area and also on service life time of the steel structures. The identification of the residual stress distribution is one of the important parts of the analysis for the evaluation of residual life time of steel structures and estimation of critical defect size of the structure. Stresses are responsible for local deformation of weld area and influence properties and behaviour of these structures during service. The initiation of reheat and stress corrosion cracking limit the life time of steel structures. Combined effect of material quality, environment and level of stresses, residual and/or external, play an important role. Cracks/defects that are larger than the tolerated limit have to be either removed or repaired. This approach is used for storage tanks used for holding of anhydrous ammonia, LPG, NGL, gasoline, naphtha, butadiene, ethylene, hydrogen, oxygen, nitrogen, argon, LNG (liquefied natural gas), biogas, sewage gas and waste water.

Previous neutron diffraction measurement of stress distribution in spherical storage tank butt weld made of C-Mn unalloyed steel, measured at three different distances (3.5 mm, 7.5 mm and 11.5 mm) below the specimen's surface, demonstrated that the highest tensile stresses are located close to the weld root surface in the weld metal [1]. This contribution presents the results of strain/stress

[a] Corresponding author : hervoches@ujf.cas.cz

distribution measurements, obtained by the neutron diffraction method, in the close vicinity of the butt weld from the spherical storage tank.

2 Experimental

The investigated storage tank material has is made of C-Mn unalloyed steel, strength of approx. 430 MPa and thickness of 30 mm. A specimen of dimensions of 150 x 200 x 15 mm was prepared for the measurements. Normal, transverse and longitudinal components have been measured for the through thickness positions 2 mm below the surface, on the neutron diffraction strain/stress scanner SPN-100 [2, 3] installed at the research reactor LVR-15 in Řež [4]. The instrument is equipped with a focusing Si(111) monochromator and a 2D position sensitive detector (Figure 1). The specimen was attached to a six-axis robotic arm for positioning in the appropriate orientations for measurement of the normal, transverse and longitudinal strain components measurements (Figure 2).

The measurements were done at a fixed position of the 2D detector with the centre of the detector positioned at an angle of $2\theta = 65°$ and an angular range of 12 degrees, allowing the observation of the (110) crystallographic plane of the ferrite, α-Fe(110). The constant neutron wavelength was 0.213 nm. A sample gauge volume of 3 mm ×3 mm ×3 mm was defined using fixed cadmium slits in the incident and diffracted neutron beam.

The strain determination is based on measurements of angular deviation $\Delta\theta_{hkl}$ of the diffraction profile position from the value related to the stress-free state. The differentiated Bragg's equation gives a relative lattice strain ε_{hkl} according to the formula:

$$\varepsilon = \frac{\Delta d}{d_0} = \frac{\sin \Theta_0}{\sin \Theta} - 1 \tag{1}$$

Where d_0 is the stress-free interplanar spacing.

The stress components σx, σy, and σz are consequently obtained using Hooke's law:

$$\sigma_x = \frac{E}{(1-2v)(1+v)}[(1-v)\varepsilon_x + v(\varepsilon_y + \varepsilon_z)] \tag{2}$$

Where $\varepsilon_{x,y,z}$ is the x-, y-, z-component of the lattice strain measured on the crystal lattice planes (hkl), E the diffraction elastic Young's modulus, and v the diffraction Poisson's ratio of the considered material.

For α-Fe(110), Young's modulus E = 225.5 GPa, and Poisson's ratio $v = 0.28$ values were used. The SteCa software [5] was used to extract 2θ scattering angle and FWHM of the diffraction peaks from the recorded 2D diffraction images.

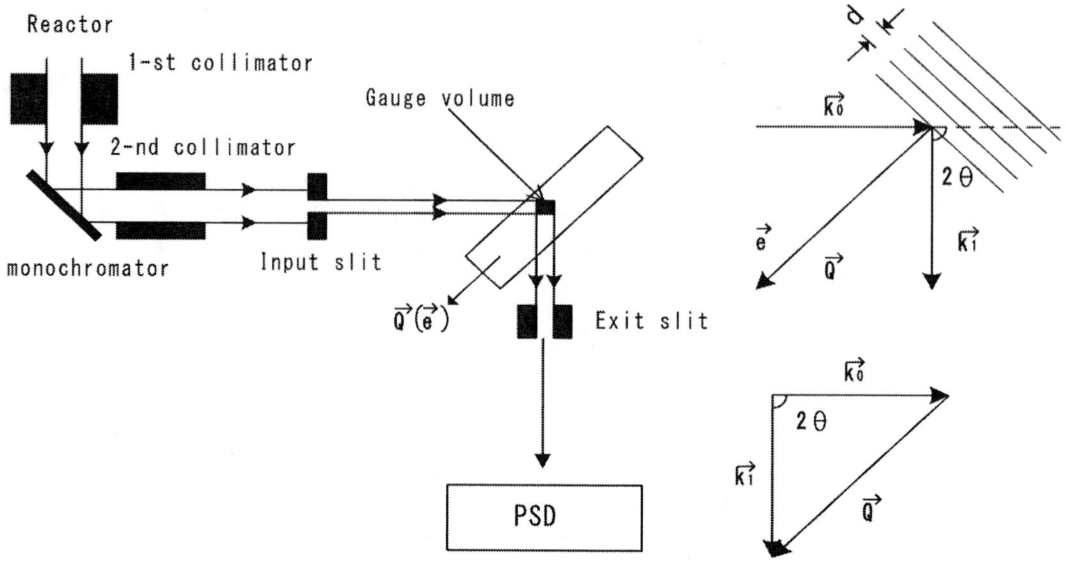

Figure 1. Schematic sketch of the neutron strain/stress scanner: Q –scattering vector perpendicular to the lattice planes, PSD – position sensitive detector, d – lattice distance, $k_{0,1}$ - neutron wave vector, ϑ - Bragg angle.

3 Results and Discussion

The welded joint made of unalloyed C-Mn steel yield strength of 255 MPa and tensile strength of 410 MPa has been used for identification of residual stress distribution. Two plates A and B thickness of 30 mm have been used to create vertical weld of spherical pressure vessel. Chemical composition of both plates is summarized in Table 1. Chemical composition of weld metal is shown in Table 2. Manual metal arc welding process using E-Boi spec. electrodes has been employed for the preparation of the weld. Macrostructure of the analysed welded joint is shown in Figure 3.

Table 1. Chemical composition of C-Mn steel used for spherical vessel.

plate	Mass concentration [%]									
	C	**Si**	**Mn**	**P**	**S**	**Cu**	**Ni**	**Al**	**Cr**	**Mo**
A	0,158	0,291	0,542	0,019	0,016	0,064	0,065	0,050	0,124	0,022
B	0,185	0,246	0,405	0,017	0,012	0,056	0,073	0,036	0,148	0,056
11 418.1 according to STN 41 1418	max. 0,20	max. 0,35	0,50 1,40	max. 0,030	max. 0,025	max. 0,30	max. 0,30	min. 0,020	max. 0,25	max. 0,08

Table 2. Chemical composition of weld metal.

Mass concentration [%]					
C	**Mn**	**Si**	**P**	**S**	**Mo**
0,090	0,587	0,257	0,023	0,012	0,410

Figure 2. Photograph of the sample attached to a robotic arm, for neutron diffraction.

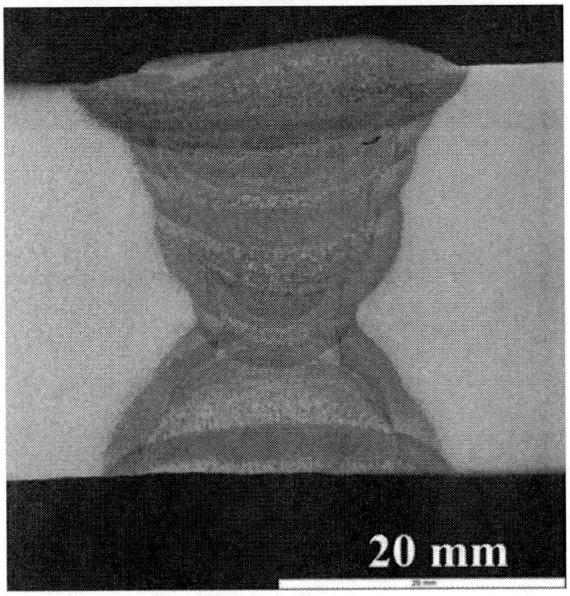

Figure 3. Macrostructure of the analysed welded joint.

Normal, transverse and longitudinal components have been measured for the through thickness positions 2 mm below the surface on the neutron diffraction. The relatively large defined sample gauge volume (3 mm ×3 mm ×3 mm) used for measurement close to the sample surface could induce pseudo-strains, due to the so-called surface effect. Several methods able to deal with this well-known phenomenon have been developed [6], often requiring extra measurements time. A recently described semi-empirical method requiring no extra calibration measurements could also be applied [7].

The measurements completed with the unalloyed C-Mn steel material revealed that maximum residual stresses in the level of 300 MPa are observed in the longitudinal direction, 150 MPa in transversal and 200 MPa in normal direction (perpendicular to the main face of the plate), as shown in Figure 4.

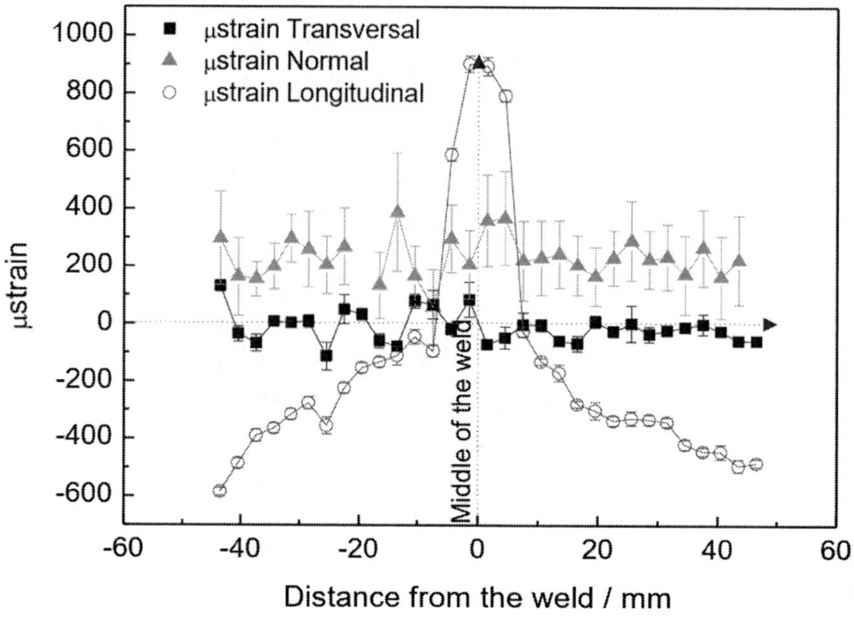

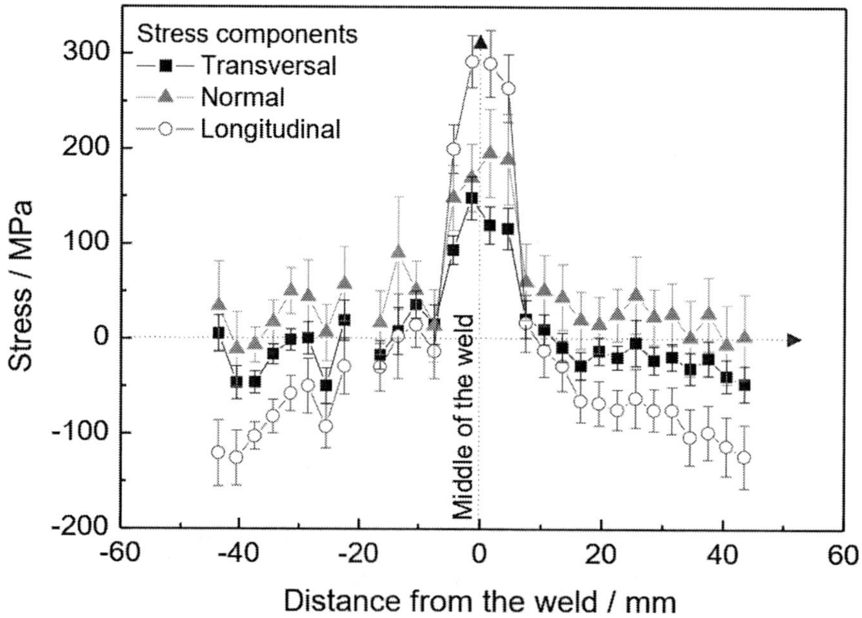

Figure 4. Microstrains and stresses versus the distance from the weld obtained by neutron diffraction.

4 Conclusions

Residual stress distribution in the vicinity of the weld taken out of spherical tank used for storage of anhydrous ammonia has been determined using neutron diffraction. The measured tensile stresses, 2 mm below the sample surface, with a value of ~300 MPa, are higher than the residual stresses measured deeper in the sample. Consequently, the results were used for calculation of the critical crack size in order to take decision concerning the necessary repair of the pressure vessel.

Acknowledgements

This work was supported by the Czech Science Foundation project No.16-08803J. Measurements were carried out with the support of the Ministry of Education, Youth and Sports (CANAM infrastructure of the NPI CAS Řež - project No. LM2015056, and infrastructure "Reactors LVR-15 and LR-0" - project No. LM2015074).

References

1. L. Mráz, C. Hervoches, M. Vrána, P. Mikula, P. Baksa, and J. Bošanský, *EAN 2017 - 55th Conference on Experimental Stress Analysis*, 133-138 (2017).
2. P. Mikula, M. Vrana, P. Lukas, J. Saroun, P. Strunz, H.J. Ullrich and V. Wagner, *Proc. of ICRS-5, T. Ericsson, M. Odén and A. Andersson (Eds.)* **2**, 721-725 (1997).
3. Ch.H. Hervoches, P. Mikula, M. Vrána, *EAN 2015 - 53rd Conference on Experimental Stress Analysis*, 119-120 (2015).
4. P. Mikula, Materials Structure **13**, 51-62 (2006).
5. C. Randau, U. Garbe, and H. G. Brokmeier, J. Appl. Cryst. **44**, 641–646 (2011).
6. M.T. Hutchings, P.J. Withers, T.M. Holden, T. Lorentzen, *Introduction to the Characterization of Residual Stresses by Neutron Diffraction*, CRC Press, Taylor and Francis, London (2005).
7. J. Šaroun, J. Rebelo-Kornmeier, J. Gibmeier, M. Hofmann, Physica B: Condensed Matter, *in press*, https://doi.org/10.1016/j.physb.2018.01.013 (2018).

MATEC Web of Conferences **253**, 01006 (2019) https://doi.org/10.1051/matecconf/201925301006
MSME 2018

The Research on Shear Properties of Deformable Ground Soil under High-speed Driving Conditions

LYU Wei-wei[1,a], LI Zhong-xin and LOU Peng
[1]*School of Mechanical Engineering, Beijing Institute of Technology, Beijing 100081, China*

Abstract: The classical ground soil shear mechanics model is difficult to predict the maneuverability of vehicles under high-speed driving conditions. The shear mechanical properties of soil under dynamic loading are the key factor to research on the ground attachment characteristics of vehicles under high-speed driving conditions. The relationship between the shear properties of the ground soil and the loading rate was analyzed by numerical simulation method. Based on the Janosi shear model of the ground soil, the loading rate and shear rate were supplemented to establish the applicable driving conditions for high-speed and heavy-duty vehicles. The ground soil shearing characteristics model provides a theoretical basis for researching on the vehicle's driving maneuverability under high-speed driving conditions.

1 Janosi shear model for classic ground soil

There are two kinds of shear stress-deformation characteristic curves for common ground soil at low shear rate. As shown in Fig. 1, the abscissa is the shear displacement and the ordinate is the soil shear stress. One is to describe the brittle ground soil, such as compacted sand, frozen snow, and other undisturbed hard ground. The shear stress-strain curve of this kind of soil will have a maximum value, and the shear stress of the soil will be smooth after reaching the yield limit. The shear models describing the brittle soil are: Bekker shear model, JYWong shear model and so on. The other is plastic soil, such as loose sand, dry soil,and other disturbed ground soil, the shear stress-strain curve of this kind of soil has no peak of protrusion, and it always changes smoothly. The shear models describing plastic soils are: Janosi shear model, hyperbolic shear model, pure exponential shear model, etc.[1-5].

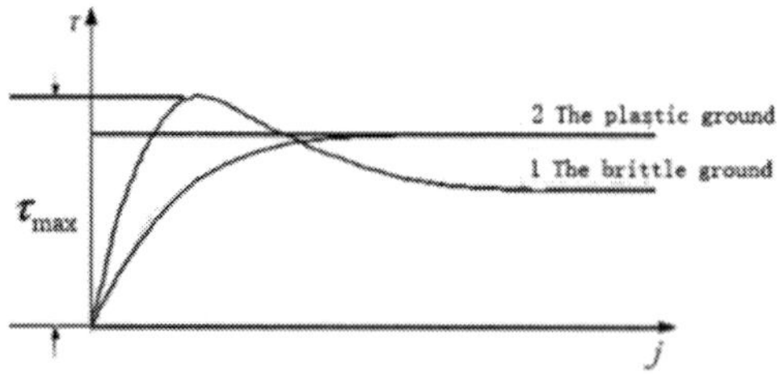

Figure 1. Two common soil shear

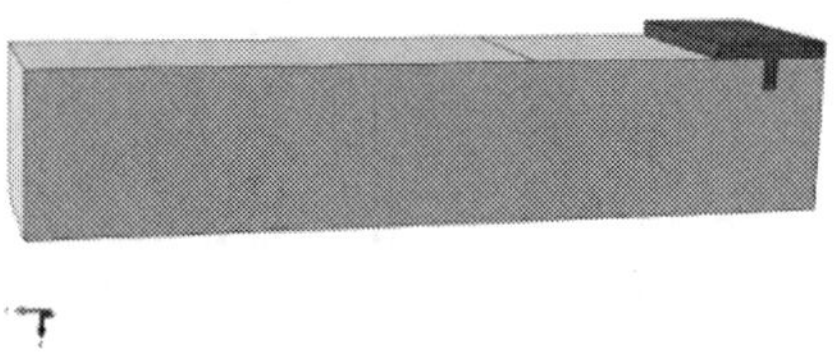

Figure 2. Simulation model of sheared stress-strain curves soil test for track shoes

[a] Corresponding author : lvweiwei@bit.edu.cn

For the plastic shear properties of deformable ground soil, there is no hump in the shear stress-strain curve, so the Janosi shear model is adopted[3]to describe the shear properties of ground soil at low shear rates, as follows:

$$\tau=\tau_{\max}^{0}\left(1-e^{-\frac{j}{k_m}}\right) \tag{1}$$

$$\tau_{\max}^{0}=c+p\tan\varphi \tag{2}$$

Where, j is shear displacement;k_m is shear modulus; $\tau_{\max}^{0}$ is soil shear strength at low shear rates; τ is shear stress; c is soil cohesion; φ is the internal friction angle of the soil; p is soil normal pressure.

2 Analysis of soil shear characteristics

2.1 Soil shear characteristics simulation model

(1)Numerical calculation model.Simulation model of the shear plate shear test in Fig. 2. Due to the limitation of indoor soil trough size, in order to reduce the influence of the shearing force of the side of the track shoe and the soil on the soil shear characteristics, the shortest side of the soil model is consistent with the short side width of the track plate. The structural parameters of the track shoes are 200 mm × 150 mm × 15 mm, and the structural parameters of the spurs are 150 mm × 10 mm × 8 mm. When researching on the shearing soil test of track shoes under high shear rate conditions, the calculated soil sizes for the track shoes and the ground are: 495mm×165mm×60mm and 165 mm×1000mm×200mm [2-7].

(2)Selection of model materials and unit types. The ABAQUS finite element software was used to establish the crawler plate and deformable ground soil simulation model. The track shoes, pressure plate and deformable ground unit model were selected as C3D4 solid elements. The material of track shoes and pressure plate was defined as rigid body, and the soil material was selected as elastoplastic Drucker-Prager constitutive material. The parameters defined for the track shoes, the press plates and the Kangzhuang soil material model with a moisture content of 5% are shown in Tables 1 and 2 [6-7].

Table 1 Material model parameters of track shoes and pressure plates

Density ρ / g · mm^{-3}	Elastic Modulus E / kPa	Poisson ratio μ
7.85×10^{-3}	2.05×10^{6}	0.3

Table 2 Soil Drucker-Prage elastoplastic constitutive model parameters

Confining pressure σ_3 /kpa	density ρ / g·mm^{-3}	Elastic Modulus E /kpa	Internal friction angle β /rad	k_1	Dilatancy angle ψ /rad	Yield stress σ_c^0 / kpa	Plastic strain $d\bar{\varepsilon}^{pl}$	Poisson ratio μ
100	1.82×10^{-3}	396	0.99	0.778	0	49	0	0.3

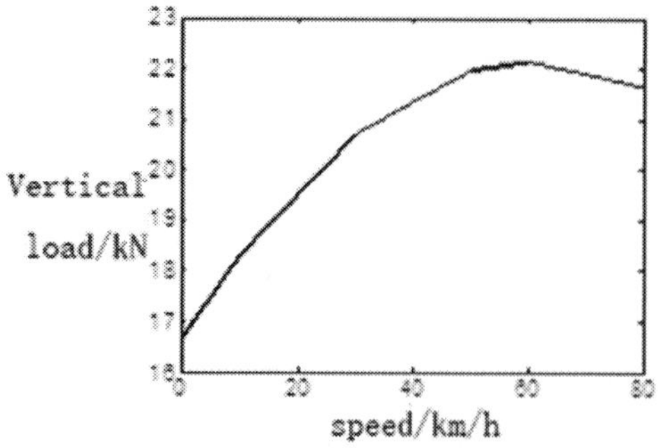

Figure 3 Relationship between driving speed and vertical load of vehicle to ground

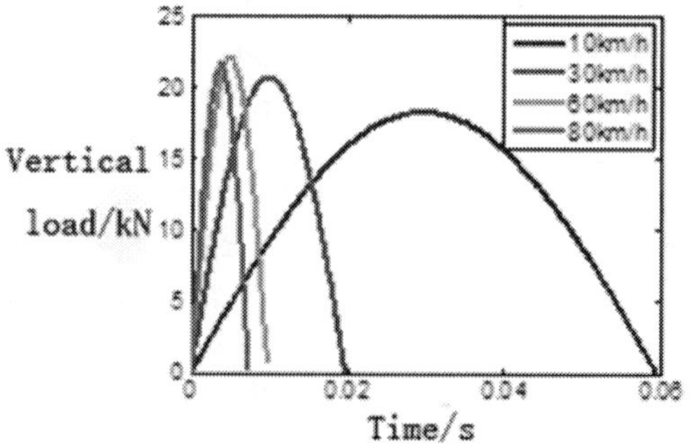

Figure 4 Semi-sinusoidal loading curve at four vehicle speeds

(3) Boundary conditions. The pressure characteristics of the deformable ground soil are studied, and the degrees of freedom in the three directions of the bottom surface of the ground soil model in Fig. 2 are all constrained, and the load or displacement in the direction of the vertical contact surface is applied to the pressure plate in Fig. 3 and Fig. 4, and all the degrees of freedom of the pressure plate except the Z direction are constrained.

2.2Influence of shear rate and water content on soil shear strength

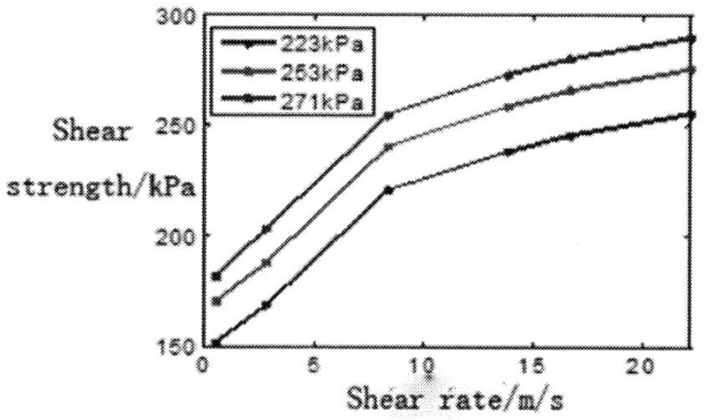

Figure 5 Relationship between soil shear strength and different normal pressures

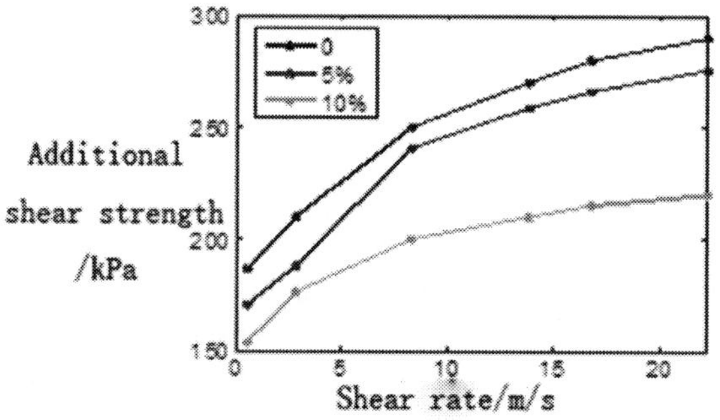

Figure 6 Relationship between shear rate and shear rate of soil with different water content

Figure 5 shows the relationship between soil shear strength and shear rate under different normal pressures. The vehicle's driving speed is 10km/h, 30km/h, 60km/h. The normal compressive stress of the track plate on the deformable ground soil is 223 kPa, 253 kPa, 271 kPa, the soil shear strength begins to change with the shear rate, and then tends to a fixed value. As shown in Fig. 5, the normal pressure of the track shoes is 223 kPa, the horizontal displacement is 150 mm, and the moving speeds are 0.2 m/s, 0.5 m/s, 2.78 m/s, 8.3 m/s, 13.8 m/s, 16.7m/s and 22.2 m/s. Relationship between soil shear strength and shear rate with water content of 0, 5% and 10% respectively. The three curves show the variation of soil shear strength and shear rate of three different water contents. The soil with higher water content increases with the increase of shear rate, and it is easier to reach a stable state. The green curve in Figure 6 indicates the water content of 10 % soil, as the shear rate increases, its shear strength changes quickly to level and increases no longer[8-11].

2.3 Shear rate amplification factor

The variation law of shear rate of ground soil under high-speed conditions was studied. A parameter reflecting the influence of shear rate on shear strength was proposed, that is shear magnification factor

η . η is the ratio of shear strength at different shear rates to soil shear strength at low shear rates, with magnification factor η is for the ordinate, the shear rate is for the abscissa.

Therefore, the relationship between the amplification factor and the shear rate can be expressed as:

$$\frac{v_2}{\eta} = a_2 + b_2 v_2 \tag{3}$$

$$\eta = \frac{\tau_{max}}{\tau^0_{max}} \tag{4}$$

where, v_2 is the rate of movement of the track shoe relative to the ground, ie the shear rate; η is magnification factor, dimensionless; a_2 and b_2 are the intercept and slope of the extension line and the ordinate axis of the fitted straight line, respectively; τ^0_{max} is soil shear strength at low shear rates; τ_{max} is soil shear strength at high shear rates.

(1) The variation of shear strength and shear rate under different normal pressures. It can be seen from Fig. 7 that the soil shear strength amplification factor and the shear rate change trend are similarunder different normal pressures. The larger the normal pressure, the soil is compacted, the internal pores are further reduced, and the soil shear amplification factor changes. The smaller the difference, the lower the normal pressure amplification factor and the shear rate change are rather the higher the normal pressure.

The shear rate is always greater than 0, the ratio of the shear rate to the magnification factor is the ordinate, the shear rate is the abscissa, and the magnification coefficient and the shear rate curve are transformed into coordinates to obtain three nearly coincident lines (as shown in Figure 7). When the shear rate is 0.5m/s, the soil shear strength amplification coefficient is 1under different normal pressures, and 3 straight lines intersect with one point. Different normal pressures influence the intercept on the straight line and the ordinate axis a_2, and the slope of the lineb_2. The greater the normal pressure, the slope of the lineb_2 is larger, the intercept on the line and the ordinate axisa_2 is smaller.

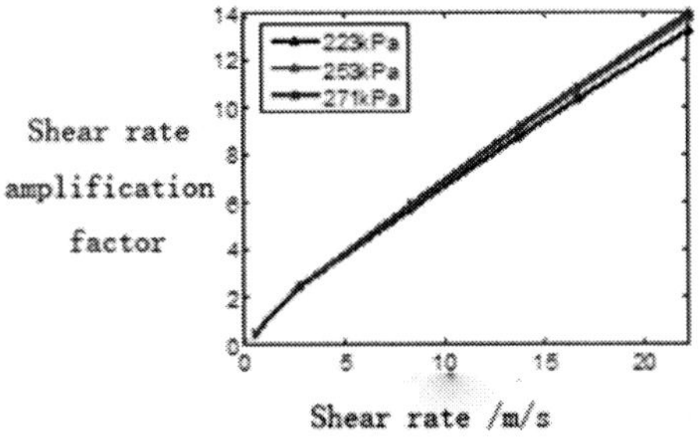

Figure 7 Relationship between different normalpressure andshear rateamplification factor

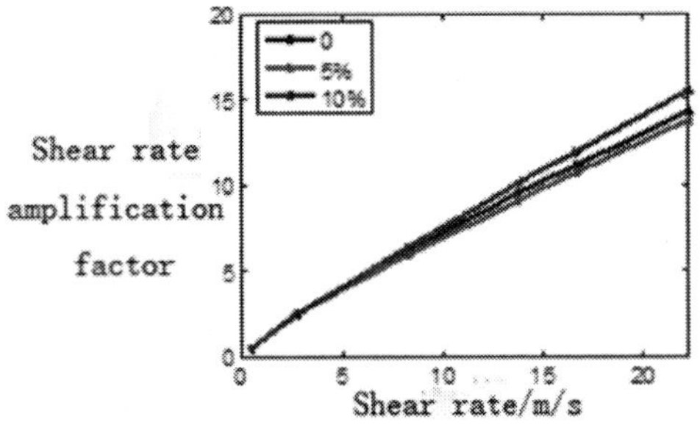

Figure 8 Relationship between different water and shear rateamplification factor

(2) Effect of differentmoisture content on soil shear strength

The three straight lines indicate the relationship between the shear rate and the amplification factor ratio of the three water contents soils and the shear ratein Fig. 8. The higher the water content, the larger the slope of the line. The three straight lines intersect with one point, which indicates the soil shear strength at a low shear rate of 0.2 m/s, where the soil strength amplification factor is 1, so that the three straight lines intersect at the same point in the coordinate system. Different water content and different normal pressureinfluence on the intercept and straight line slope of the linear extension line and the ordinate, indicating t the relationship between shear rate and soil shear strength under different normal pressures and soil moisture content can be described by parameters a_2 and b_2 .

2.4 Influence of soil water content and normal pressure on model parameters a_2 and b_2

The research found that the soil strength amplification factor is directly related to soil water content and normal pressure. The parameters obtained by analyzing the three water contents, three normal pressure shear rates and the shear rate curve are shown in Fig. 9 and Fig. 10 The relationship between rate and normal pressure and parameter sum.

Soil strength factorIt is directly related to soil water content and normal pressure. The parameters obtained by analyzing three kinds of water content, three kinds of normal pressure shear rate and shear rate curve are analyzed.a_2 withb_2 Figure 9 and Figure 10 show soil moisture content and normal pressure and parameters. $a_2$with b_2 Relationship.

The red dots indicate the parameters a_2 and b_2 obtained for different moisture content and normal pressure in Figures 9 and 10. The moisture content has a great influence on the parameter a_2 and b_2. The moisture content increase, parameter a_2 decreases rapidly and the parameters b_2 gradually increase; the normal pressure has little effect on the parameters a_2. In Fig. 9, the normal pressure increases and the parameters has not changed much at the same moisture content. In Figure 10, when the normal pressure increases, the parameters b_2 show a trend of increasing first and then decreasingat the same moisture content.

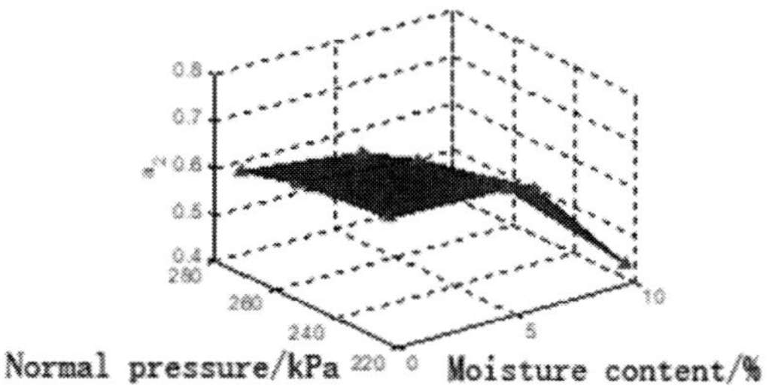

Figure 9 Influence of moisture content and normal pressure on parameters a_2

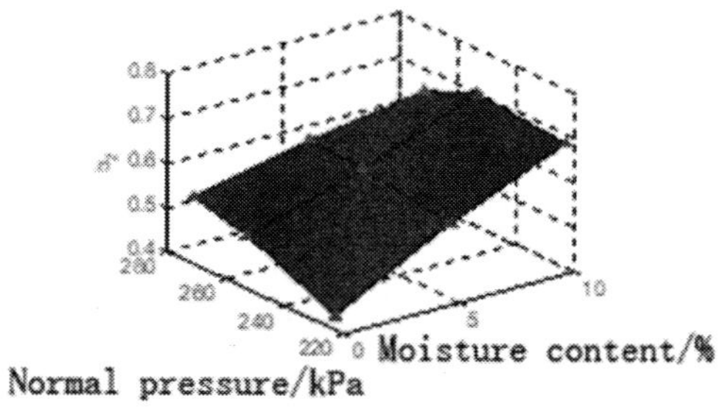

Figure 10 Influence of moisture content and normal pressure on parameters b_2

3 shear correction model

The shear deformation of deformable ground soil is affected by the shear rate of the track shoes and the vehicle's load on the ground. Under the same load, the shear rate can increase the strength of the soil. Based on the Coulomb model [12-15], shear is introduced. Intensity amplification factor, establish a soil strength formula related to driving speed.

$$\tau=\tau_{\max}\left(1-e^{-\frac{j}{k_m}}\right)=\eta\tau_{\max}^{0}\left(1-e^{-\frac{j}{k_m}}\right)=\frac{v_2}{a_2+b_2v_2}(c+p\tan\varphi)\left(1-e^{-\frac{j}{k_m}}\right) \tag{5}$$

Where, jis shear displacement; kmisshear modulus; $\tau_{\max}^{0}$ is soil shear strength at low shear rates; τ is shear stress; c is soil cohesion; φ is the internal friction angle of the soil; p is soil normal pressure; v2 is the rate of movement of the track shoe relative to the ground, is the shear rate ($v \geq 0.2$m/s); η is soil strength amplification factor, dimensionless; a_2 and b_2 are the intercept and slope of the extension line and the ordinate axis of the fitted straight line, respectively; $\tau_{\max}$ is soil shear strength at high shear rates.

Table 3 Model parameters before and after soil shear model correction

Model parameter	Before correction	After correction
Moisture content w / %	4	4
Cohesion c / kPa	17.82	17.82
Internal friction angle φ /°	33.7	33.7
Shear rate v_2 / m/s	10	10
Normal pressure p / kPa	223	223
Shear deformation modulus k_m /mm	12	10
a_2 / m/s	0	0.66
b_2	0	0.5

Using the parameters in Table 3, using the modified shear model for calculation, comparing the calculation results before the correction with the numerical simulation results, which shows that the corrected model has a better prediction effect, and the error with the numerical calculation result is up to 25%.

References

1. Wong J. Y., Huang W. "Wheels vs. tracks"A fundamental evaluation from the traction perspective[J]. Journal of Terramechanics, 2006, 43(1): 27-42.
2. Zhuang Jide, Calculating Vehicle Ground Mechanics [M]: Mechanical Industry Press, 2002.
3. Zhang Keyuan, Vehicle Ground Mechanics [M]: Beijing: National Defense Industry Press, 2002.
4. Park W. Y., Chang Y. C., Lee S. S.et al. Prediction of the tractive performance of a flexible tracked vehicle[J]. Journal of Terramechanics, 2008, 45(1): 13-23.
5. Liu K., Ayers P., Howard H.et al. Influence of soil and vehicle parameters on soil rut formation[J]. Journal of Terramechanics, 2010, 47(3): 143-150.
6. Al-Milli S., Seneviratne L. D., Althoefer K. Track--terrain modelling and traversability prediction fortracked vehicles on soft terrain[J]. Journal of Terramechanics, 2010, 47(3): 151-160.
7. Lyasko M. LSA model for sinkage predictions[J]. Journal of Terramechanics, 2010, 47(1): 1-19.
8. Lyasko M. Slip sinkage effect in soil--vehicle mechanics[J]. Journal of Terramechanics, 2010, 47(1): 21-31.
9. Lyasko M. Multi-pass effect on off-road vehicle tractive performance[J]. Journal of Terramechanics, 2010, 47(5): 275-294.
10. Lyasko M. I. How to calculate the effect of soil conditions on tractive performance[J]. Journal of Terramechanics, 2010, 47(6): 423-445.
11. Alaoui A., Diserens E. Changes in soil structure following passage of a tracked heavy machine[J]. Geoderma, 2011, 163(3): 283-290.
12. Abo-Elnor M., Hamilton R., Boyle J. T. 3D Dynamic analysis of soil--tool interaction using the finite element method[J]. Journal of terramechanics, 2003, 40(1): 51-62.
13. Lee J. H. Finite element modeling of interfacial forces and contact stresses of pneumatic tire on fresh snow for combined longitudinal and lateral slips[J]. Journal of Terramechanics, 2011, 48(3): 171-197.
14. Hambleton J. P., Drescher A. Modeling wheel-induced rutting in soils: Rolling[J]. Journal of Terramechanics, 2009, 46(2): 35-47.
15. Hambleton J. P., Drescher A. Modeling wheel-induced rutting in soils: indentation[J]. Journal of Terramechanics, 2008, 45(6): 201-211.

The Effect of Hypervelocity Impact Loading on Explosively Welded Ti/Al/Al Plate

Marcin Wachowski[1,a], Teresa Fras[2], Robert Kosturek[1], Lucjan Śnieżek[1], Ireneusz Szachogłuchowicz[1] and Krzysztof Grzelak[1]

[1] *Military University of Technology, Faculty of Mechanical Engineering, 2 gen. W. Urbanowicza str., 00-908 Warsaw, Poland*
[2] *French-German Research Institute of Saint-Louis (ISL), 5 rue du General Cassagnou, 68301 Saint-Louis, France*

Abstract. The aim of this paper is to investigate the ballistic properties of explosively welded Ti6Al4V/AA1050/AA2519 laminate under hypervelocity (4000 m/s) impact loading of projectile made of aluminium ball. The paper describes the influence of the projectile impact energy on the structure of the laminate components and delamination of bonds depending on its strength and presence of intermetallic phases. Observation of the failure is performed for the plate impacted for aluminium alloy side. Results revealed ductile shearing as a dominant process leading to perforation.

1 Introduction

One of the most promising engineering materials in terms of the improvement ballistic properties of shields used in space environmental constructions are metallic laminates [1-3]. The main problem of the construction stays in space is presence of debris, cosmic dust and trashes e.g. parts of old satellites which travel with very high speed often exceed thousands of meters per second. These elements constitute main threat for space stations and satellites because potential collisions can cause significant damage of construction. Currently applied shields protecting against debris consist of several very thick layers of aluminium alloy. Additionally, each layer is separated to reduce the fragmentation phenomenon. The approach that is worth consideration is using laminates to provide the necessary ballistic properties with ensured the lowest weight and thickness of the shield used in space constructions. One of the most interesting metallic laminate which fully satisfies these requirements is aluminium alloy AA2519 – titanium alloy Ti6Al4V laminate. The appropriate technology to clad aluminium alloy with high strength titanium alloy is explosive welding method [4-7]. To obtain explosively welded joint between Ti6Al4V and AA2519 applying additional interlayer made of aluminium alloy AA1050 was necessary. Explosive welding is a solid-state welding technology which allows to create a metallic bond between two metal surfaces by the energy of detonation of high explosive material [8-11]. Harsh conditions of the process influence on bond zone and base materials causing their hardening and present of intermetallic phases in the joint. Presence of very hard and brittle intermetallic zones adjacent to the joint significantly increase of ballistic properties of the whole laminate [12-15]. The presented paper discusses ballistic properties of the explosively welded

[a] Corresponding author : marcin.wachowski@wat.edu.pl

Ti6Al4V/A1050/AA2519 plates impacted by aluminium ball accelerated to hypervelocity equal 4000 m/s. Investigated laminate is considered as an armour in space application. In the performed investigation aluminium ball projectile accelerated to hypervelocity condition is equivalent of debris flying in space which constituting threat for construction in space e.g. satellites, space stations. The microstructure observation and fractography investigation of perforation zone using light microscopy (LM) and scanning electron microscopy (SEM) allows to understand mechanisms of penetration and perforation in explosively-welded layered structures under hypervelocity impact loadings.

2 Experimental part

The titanium alloy Ti6Al4V has been successfully cladded with aluminium alloy 1050 and aluminium alloy 2519 by using explosive welding process. As an explosive material, the mixture of ammonium nitrate fuel oil (ANFO) was used. The scheme of the explosive welding system is shown in Figure 1. The thickness of the welded elements was equal to 5 mm for Ti6Al4V, 5 mm for AA2519 and 1 mm for AA1050. Before explosive welding the joined materials are cleaned, degreased and oxides are removed. Explosive welding process was performed by EXPLOMET High-Energy Techniques Works.

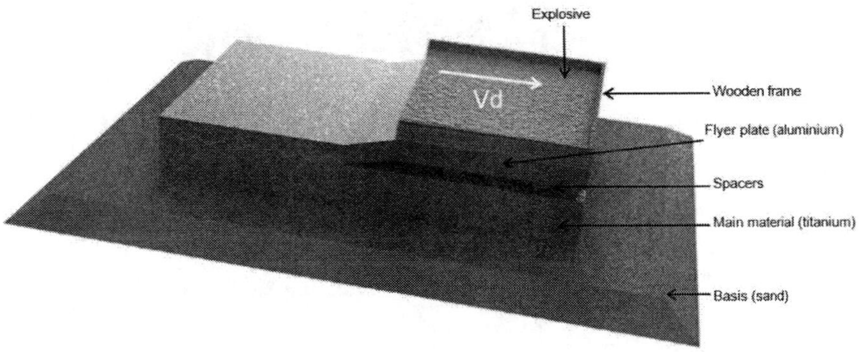

Figure 1. Scheme of explosive welding process. Vd – velocity of explosive detonation.

3 Results and discussion

3.1 Microstructure

The explosive welding process allowed to obtain high-quality Ti6Al4V/AA1050/AA2519 joint. Investigated laminate is characterized by the presence of intermetallic phases in the melted zone adjacent to the Ti6Al4V/AA1050 joint (Figure 2). Scanning electron microscopy (SEM) observation revealed flat-shaped joint with lack of delamination and voids. Presence of intermetallic phases and melted zones is the result of the high velocity, high temperature and high pressure generated during collision. SEM observation of AA1050/AA2519 joint revealed characteristic wavy-shape joint with no imperfections, such as cracks and voids. This joint is also characterized by the lack of intermetallic phases. Discrepancy in shape of joints is the results of differences in density and mechanical properties of the joined materials. In the melted zones, the investigation revealed the presence of fragments of Ti6Al4V surface layer, which underwent partial fragmentation during welding process in this area, as evidenced by analysis of titanium and aluminium distribution at the sample surface (Figure 3, Table 1).

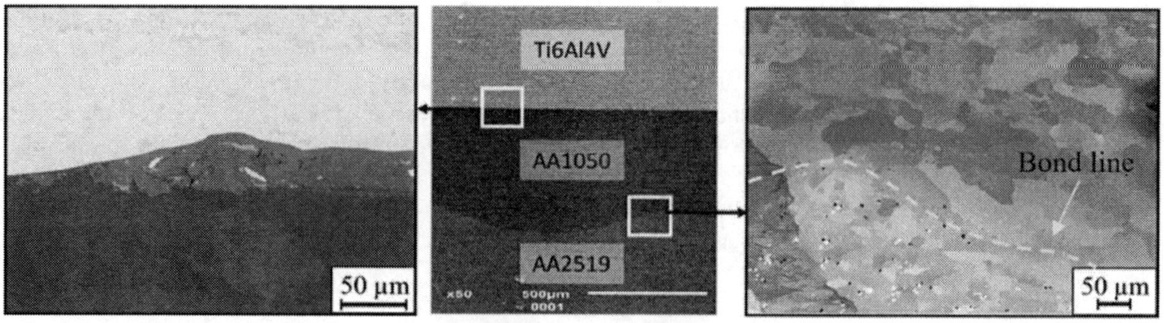

Figure 2. Microstructure of the laminate components.

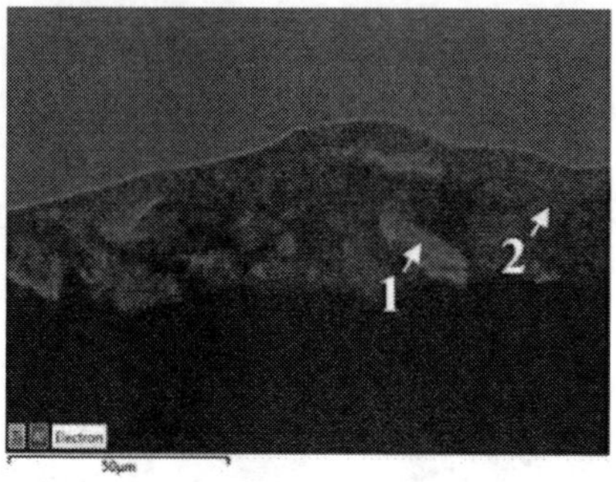

Figure 3. Microstructure of the Ti6Al4V/AA1050 joint.

Table 1. EDX mapping of the Ti6Al4V/AA1050 joint.

Point	Chemical composition % weight	
	Al	Ti
1	6.5	93.5
2	64.4	35.6

3.2 Hypervelocity Ballistic Impact Test

The cross-sections of the sample revealed that the projectile impact caused a large deformation of the plates and a delamination between AA1050 aluminium and Ti6Al4V layers. Ballistic research results revealed penetration of the projectile through AA2519 plate. Diameter of the resulting crater was about 55 mm. SEM images of the titanium and aluminium (Figure 4) depict smooth surfaces, which reveal that the ductile shearing is as a dominant process leading to perforation. In the first impact stage, a part of the energy of the projectile was acquired by the AA2519 plate. The result of this process is the dispersion and deceleration of the shock wave. The slower spread of impact energy in the material causes the formation of the crater neck due to frontal petalling process of AA2519 alloy. During the penetration, the moving projectile pushed the structure directly below its nose, which caused stretching of the laminate layers. In Ti6Al4V/AA1050 joint has been delaminated during the shooting (Figure 5). Delamination indicate that the material has reached the maximum strength limit due to perforation with the projectile and is a result of absorption of the generated energy due to the partial reflection of the shock wave from the titanium alloy. This effect minimizes the phenomenon of generating debris in the space due to the plastic deformation of the material on a larger surface. Delamination, therefore, is a phenomenon that positively affects the improvement of the ballistic properties of the produced laminate. A phenomenon favouring the delamination of Ti6Al4V/AA1050 joint is the process of partial reflection of the shock wave at the boundary of materials with two different densities. As a result, the primary shock wave loads the material towards the titanium and the reflected wave causes stretch of the joint in the opposite direction. Therefore, the joint is stretched in two opposite directions, what causes the occurrence of high shear stresses in the joint surface.

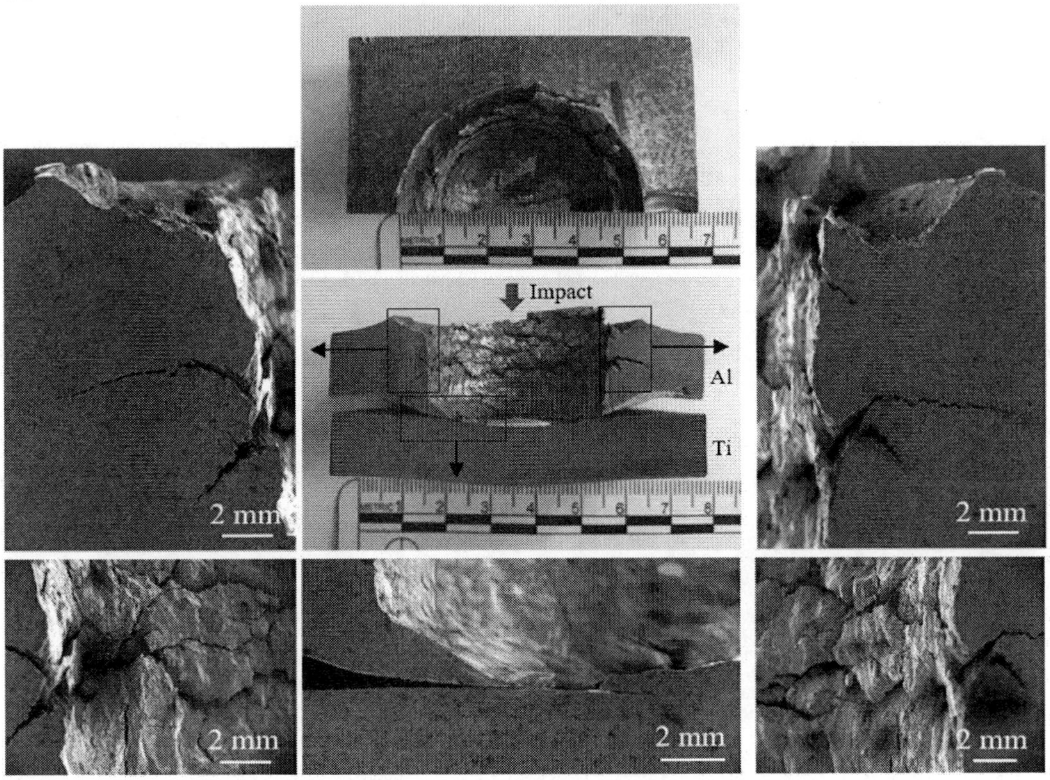

Figure 4. Fractography observation of the specimen after hypervelocity impact test.

The performed investigations showed that delamination occurs only in the Ti6Al4V/AA1050 joint (Figure 5A) and in case of AA1050/AA2519 joint despite the severe deformation due to ballistic load the cohesion of the materials bound is preserved (Figure 5B). The main reason of this phenomenon is the fact that in contrast to Ti6Al4V/AA1050 joint the joint between two aluminum alloys is free of intermetallic phases. The Ti6Al4V/AA1050 joint surface is characterized by presence of brittle phases from Ti-Al system, which tend to crack during dynamic load and therefore cause the breaking of cohesion between two plates.

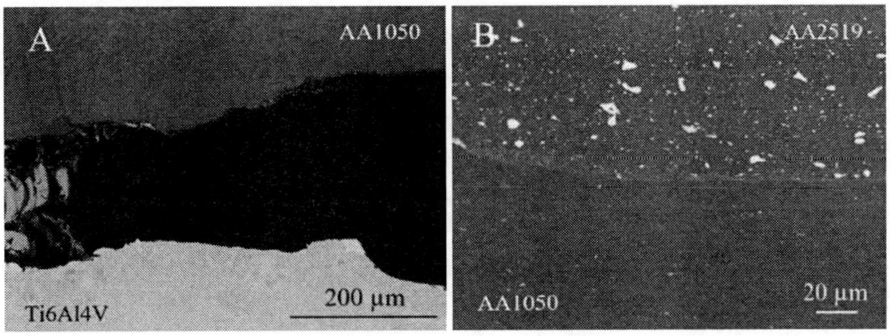

Figure 5. SEM observation of the specimen after hypervelocity impact test.
A) Ti6Al4V/AA1050 joint, B) AA1050/AA2519 joint

4 Conclusions

Explosive welding of aluminum alloy AA2519 with high strength titanium alloy Ti6Al4V using interlayer AA1050 allows to obtain a good quality joints free of cracks and voids, with the participation of melted zones and intermetallic phases between titanium alloy and aluminum AA1050. The ballistic performance of a Ti6Al4V/AA1050/AA2519 laminate was investigated in ballistic test to observe failure mechanism of material under hypervelocity conditions of impact. Result revealed that laminate is well suited to ballistic protection applications because of the combination of the material properties and structure of their constituents. The delamination, which took place on the surface of Ti6Al4V/A1050 joint has been caused by presence of the brittle intermetallic compounds localized in the melted zone on the joint interface. Fractography observation revealed that the ductile shearing is as a dominant process leading to perforation through the laminate components. Delamination of Ti6Al4V/AA1050 joint is results of the simultaneous impact of the falling impact wave and the shock wave reflected from the boundary of materials with two different densities.

Acknowledgments

This work uses results of the research from the project co-financed from Polish Ministry of National Defence No. PBG/13-998. This work was also supported by the Project PBS2/A5/35/2013 funded by the National Research and Development Centre.

References

1. I. Crouch, *The Science of Armour Materials* (Woodhead Publishing 2016).
2. H. Gower, D.S. Cronin, A. Plumtree, Inter. J. of Imp. Eng. **35**, 1000-1008 (2008).
3. M. Übeyli, Y. Orhan, B. Ögel, Mat. & Des. **28**, 1257-1262 (2007).
4. D. Boroński, M. Kotyk, P. Maćkowiak, L. Śnieżek, Mat. & Des. **133** (2017).
5. P. Bazarnik, B. Adamczyk-Cieślak, A. Gałka, B. Płonka, L. Śnieżek, M. Cantoni, M. Lewandowska, Mat. & Des. **111** (2016).
6. P. Bazarnik, M. Cantoni, L. Śnieżek, M. Lewandowska, European Microscopy Congress 2016: Proceedings, (Poland, 2016)
7. L. Śnieżek, I. Szachogluchowicz, M. Wachowski, J. Torzewski, J. Mierzynski, Proc. Str. Int. **5**, 422-429 (2017).
8. T.Z. Blazynski, *Explosive Welding, Forming and Compaction, Applied Science* (1983).
9. F. Findik, Mat. & Des. **32**, 1081-1093 (2011).
10. R. Kosturek, M. Najwer, P. Nieslony, M. Wachowski, Adv. in Man., 681-686 (2018).
11. S.A.A Akbari Mousavi, L.M. Barrett, S. Al-Hassani, J. of Mat. Proc. Tech. **202**. 224-239 (2008).
12. D.M. Fronczek, R. Chulist, L. Litynska-Dobrzynska, G.A. Lopez, A. Wierzbicka-Miernik, N. Schell, Z. Szulc, J. Wojewoda-Budka, Metall. and Mat. Trans. A **48**, 4154–4165 (2017).
13. G.H.S.F.L. Carvalho, I. Galvão, R. Mendes, R. Leal, A. Loureiro, Mat. Ch. **142**, 432-442 (2018).
14. D.M. Fronczek, R. Chulist, Z. Szulc, J. Wojewoda-Budka, Growth kinetics of TiAl3 phase in annealed Al/Ti/Al explosively welded clads, Mat. Let. **198**, 160–163 (2017).
15. B. Greenberg, M. Ivanov, M. Pushkin, A. Inozemtsev, A. Patselov, A. Tankeyev, S. Kuzmin, V. Lysak, Met. and Mat. Tran. A **47**, 5461-5473 (2016).

Volume Deformation Characteristics of Concrete Mixture during Thermal Curing Process

Y Xiang[a], Y Xie, G Long and L Li

School of Civil Engineering, Central South University, No 22 South Shaoshan Rd, 410075, Changsha, Hunan, PR China

Abstract. Expansion deformation of concrete mixture during steam curing process was a kind of unrecoverable volume expansion which would definitely affect the performance of concrete component. Because there were a few researches on expansion deformation of concrete mixture during thermal curing process, a new test method composed of constant temperature water bath, eddy current measurement sensor and data processing program was developed to measure the expansion deformation. The volume deformation characteristics of a typical concrete used for preparing track slab of China high-speed railway were investigated. Test results indicated that volume deformation of concrete mixture could be divided into 2 stages. One was rapid expansion at heating stage and another one was slow shrinkage at constant temperature stage. Besides, the addition of combination of flying ash, slag, nano-silica and nano-calcium carbonate was benefit for controlling the eventual expansion deformation during the whole thermal curing process. Study could be an important reference for heightening quality of steam-cured concrete.

1 Introduction

China is now rapidly developing the high-speed railway. In order to improve the turnover rate of mould and accelerate the construction schedule to ensure the project would be completed on time or in advance, many concrete components such as sleeper, track slab and prestressed simply supported box girder are prefabricated by means of high temperature steam curing. Then, these concrete components would be towed to construction sites for assembly and splicing after experiencing above curing process. It is true that steam-cured concrete components could be quickly produced batches; however, through years of monitoring a large number of high-speed rail projects, researchers have found that this kind of concrete exposes a considerable number of problems during the service process [1-5].

Experiments prove that due to a series of physical and chemical reactions during steam curing process happened within concrete mixture, volume expansion, chemical shrinkage, autogenous shrinkage and temperature deformation are produced which always results in causing surface and interior cracks. These initial cracks would begin to deteriorate during the long-term service under alternate cycling of temperature, rainwater, chemistry and train dynamic load conditions. Finally, mechanical and durability performances of concrete components are eventually becoming worsen and the service life is definitely shortened [6-8].

Expansion deformation is the most prominent thermal damage of concrete mixture during steam curing process. This kind of volume deformation is a comprehensive deformation which consists of

[a] Corresponding author : Niesh2013@126.com

minority of expansion of solid phase and chemical shrinkage of hydration and majority of free expansion of liquid phase and vapour phase. Since the restrain effect of mould, internal vapour moves upward drives some cementing component particles to move upward which makes the early concrete skeleton structure grow upward. With continuous hydration of cementing component particles, those products gradually fill up pores and cracks caused by vapour evaporation and the framework of concrete is basically formed, which would finally cause the unrecoverable volume deformation.

Nowadays, because there are a few researches on volume deformation characteristics of concrete mixture during steam curing process, this study has developed a new test method to imitatively measure the deformation. The influences of different cementing components on volume deformation of concrete mixtures during thermal curing process are analyzed and the deformation rule is further explained by testing evaporable water content and compressive strength. Study provides an important reference of controlling volume deformation and heightening quality of steam-cured concrete.

2 Raw materail and methodology

2.1 Raw materials

Raw materials adopted in this study include cement (C), flying ash (FA), ground granulated blast furnace slag (GGBS), nano-silica (NS), nano-calcium carbonate (NC) particles, coarse and fine aggregates, high-efficient superplasticizer (SP) and water (W). Main technical features of above raw materials are shown as followed: (I) C is P·I 42.5 Portland cement and grades of FA and GGBS are first class and S95, separately. Technical indices of C, FA, GGBS, NS and NC are presented in table 1 and table 2. (II) Fine aggregates are ordinary river sands whose fineness modulus is 2.8. Coarse aggregates are crushed limestone. (III) Water reducing efficiency of SP is 25% and tap-water is adopted to use as the mixing water.

Table 1. Technical properties of C, FA and GGBS.

Type	Content (w% by weight)								Specific area
	SiO_2	Al_2O_3	FeO	CaO	MgO	SO_2	Na_2O_{eq}	I.L	($m^2 \cdot g^{-1}$)
C	24.6	7.3	4	59.7	3.8	2.5	0.6	2.5	367
FA	52.3	26.3	9.7	3.7	1.2	0.2	1.8	5.7	450
GGBS	26.1	13.8	14.1	33.6	8.1	1.75	0.45	2.1	420

Table 2. Technical properties of nano particles.

Type	Specific area ($m^2 \cdot g^{-1}$)	Diameter (nm)	Content (w% by weight)	
			SiO_2	$CaCO_3$
NS	640±50	10±5	>99.9	\
NC	45±5	35±5	\	>96.5

2.2 Test mix proportions

Considering effects of mineral admixtures and nano particles on volume deformation characteristics of concrete mixture during thermal curing process, 6 test mix proportions of concrete specimens are designed and presented in table 3.

Table 3. Mix proportions of concrete specimens (Kg/m^3).

Type	C	FA	GGBS	NS	NC	Sand	Stone		SP	W
							5-10mm	10-20mm		
C0	450	0	0	0	0	660	484	726	4.5	135
M1	315	135	0	0	0	660	484	726	4.5	135
M2	315	0	135	0	0	660	484	726	4.5	135
M3	315	90	45	0	0	660	484	726	4.5	135
N1	313.425	90	45	0	1.575	660	484	726	4.95	135
N2	313.425	90	45	1.575	0	660	484	726	4.95	135

In above mix proportions, 30% C is replaced by FA or GGBS or both of them and another 0.5% C is further replaced by nano particles.

2.3 Test programs

2.3.1 Expansion deformation.

Expansion deformation test mainly contains following 3 procedures.

(I) Preparation for concrete specimens: 100mm×100mm×515mm size stainless steel moulds are used to form concrete specimens and there are 3 specimens in each mix proportion. A thin layer of hydraulic oil should be applied to bottom and inner walls of mould first. Beside, in order to get expansion deformation of concrete mixture under unrestrained actions, two 3mm thick cardboard shells are placed along the length direction of mould, two 10mm thick polytetrafluoroethylene boards are placed at both ends of mould and finally a layer of polytetrafluoroethylene thin film is placed inside the bottom and inner walls of mould. When reflection targets have been fixed, concrete mixture is poured layer by layer and vibrated densely. It is noteworthy that distance between 2 reflection targets is no less than 400mm. When vibration has been finished, a layer of plastic thin film should be laid on top of specimen to prevent dry shrinkage and put specimen into test room for 3h of static setting.

(II) Calibration of sensors: both cardboard shells and polytetrafluoroethylene boards should be removed first and the plastic thin film should continuously cover specimen's surface. Sensors could be connected with testing system when they have been installed and the distance between sensor and reflection target should be controlled at 1.20±0.05mm. Experimental schematic diagram and testing principle are displayed in figure 1 and equation (1).

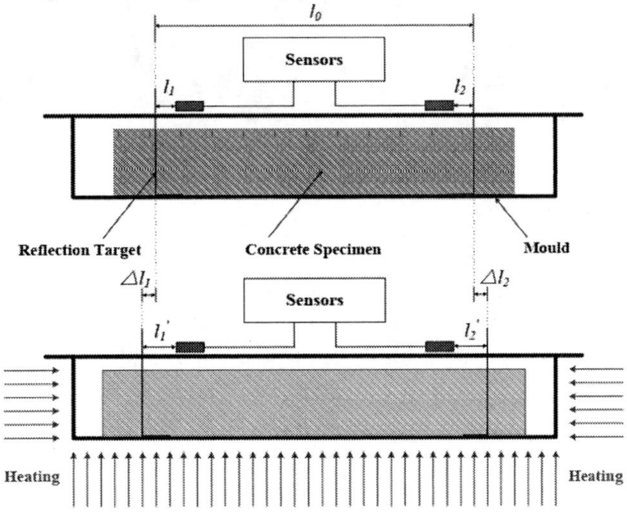

Figure 1. Experimental schematic diagram.

Calculation of volumetric expansion rate *a*:

$$\alpha = \frac{(l_1^{'} - l_1) + (l_2^{'} - l_2)}{l_0} \times 10^6 \quad (1)$$

(III) Test procedures: turning on water bath at the beginning to raise test temperature from 20°C to 60°C in 2h and then to keep the bath temperature at 60°C for 8h. During the whole thermal curing process, the data are recorded automatically once a minute by testing device and the final result is the average value of 3 specimens. Thermal curing scheme and testing device are shown in figure 2 and figure 3, respectively.

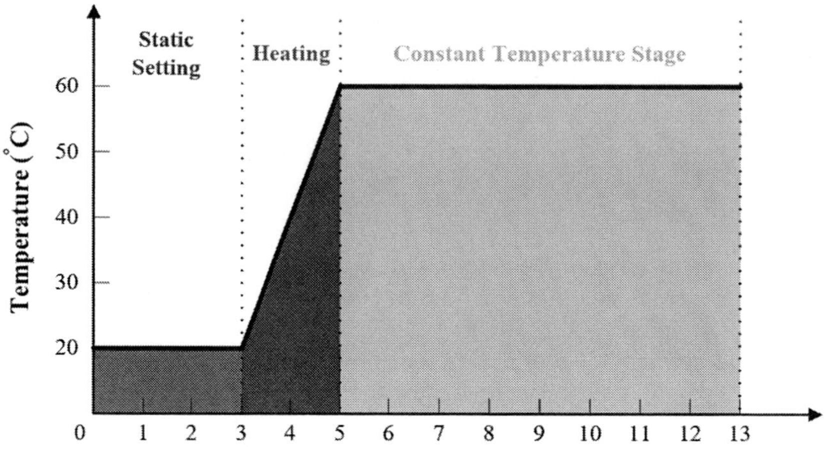

Figure 2. Thermal curing scheme.

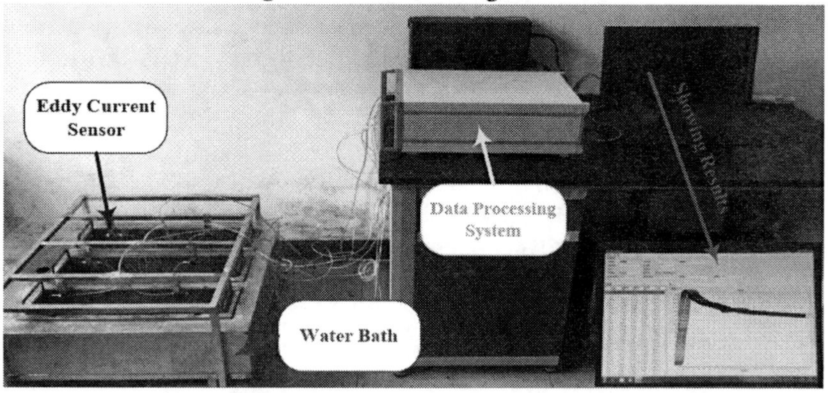

Figure 3. Experimental device.

2.3.2 Compressive strength.

Surplus concrete mixture of *2.3.1* section's test is adopted to fabricate 6 100mm×100mm×100mm size concrete cubic specimens in 2 detachable steel moulds. After the same 3h of static setting in the same test room, removing steel moulds first and compressive strengths of cubic specimens should be measured immediately. Final results would be obtained by calculating tests' average.

2.3.3 Evaporable water content.

Broken specimen of *2.3.2* section's would be further crushed into smaller pieces and then be put into a stainless steel basin whose weight is m_0 to weigh the total mass m_1. Put this basin with crushed concrete specimen pieces into an oven at 103±5°C for drying. Duing above drying process, the total mass of basin and pieces should be continuously weighted until the total mass does not change anymore and then record the last weight m_2. The evaporable water content could be calculated by following equation (2).

$$w\% = \frac{m_1 - m_2 - m_0}{m_2 - m_0} \times 100\% \tag{2}$$

3 Experimental result and dicussion

3.1 Volume deformation characteristics

It is assumed that thermal expansion coefficient of concrete mixture in the non-hardening stage is 12μ/°C while 8μ/°C in hardening stage. During thermal curing process, concrete mixture basically takes 1.5h to achieve complete hardening state. Thus, when heating stage is over, temperature strain of concrete is 440μ and this value would be kept unchanged in constant temperature stage. The true volume deformation curve of concrete is the result of test curve deducts temperature strain.

It can be seen from all volume deformation tests of concrete mixtures that volume deformation characteristics mainly contain 2 stages. On the one hand, due to insufficient hardening of concrete mixture during the temperature-rising stage, huge expansion deformation caused by water vapor's expansion makes the deformation curve rise fast. On the other hand, with the help of gradual hardening of concrete and steady temperature, expansion deformation has basically stopped. Because of continuing hydration of cementing components, a shrinkage phenomenon of concrete has came out which results in declining the deformation curve in this constant temperature stage.

Concrete mixtures incorporating different cementing components have little differences of thermal expansion ratios in temperature-rising stage; however, their thermal expansion deformation peaks are various. Effects of mineral admixtures and nano particles on volume deformation characteristics have been discussed.

3.1.1 Inhibition effects of mineral admixtures.

As is shown in figure 4, despite different thermal expansion peaks, the deformations of concrete specimens incorporating different mineral admixtures have no obvious differences in heating stage; however, when tests reach constant temperature stage, their deformation values of specimens become clearly different. Besides, shrinkage deformations are various from each specimen.

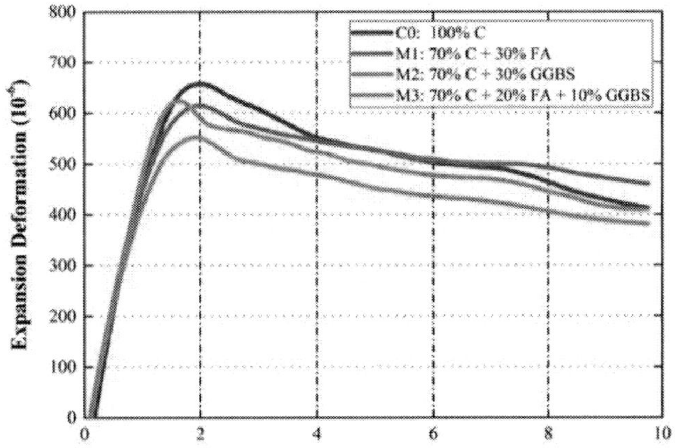

Figure 4. Effects of mineral admixtures on controlling volume deformation.

Figure 4 has presented following changes: C0 group has the biggest expansion peak 650μ but it reduces 250μ in the end. M1 and M2 groups own similar maximum expansion deformations near 615μ however M2 group could achieve that value faster. Shrinkage trend slows down when M1 group enters constant temperature stage and final expansion value is higher than C0 group whereas shrinkage of M2 group continuously enlarges and final expansion is slightly lower than C0 group. M3 group not only has the smallest expansion peak 550μ but also owns the lowest final expansion value 380μ. Above test results imply FA and GGBS have a great influence on controlling volume deformation of concrete mixture. Concrete specimen incorporating single GGBS could fast arrive at the maximum expansion peak, but concrete specimens incorporating single FA or mingle FA-GGBS could reach that peaks until heating stage is over. During the constant temperature stage, expansion deformations of concrete specimens gradually change into obvious shrinkage deformations. C0 group has the largest reduction of shrinkage deformation and followed by M2 group, which means although GGBS has replaced 30% C it still promotes hydration of cement. However, trends and values of shrinkage deformations of concrete specimens incorporating single FA or mingle FA-GGBS slightly decline. Test results present GGBS plays an important role in deformation changes of concrete at temperature-rising and constant temperature stages.

Causes of above experimental phenomenon are related to greater activity of GGBS and density of initial pore structure of concrete mixture. Though activity of FA is small and its hydration process is slow, the addition of combination of FA and GGBS still forms the "superimposed compound effect" to enhance the initial structure [9-10]. Thus, M3 group heightens compactness and initial pore structure and promotes early hydration to generate higher expansion deformation resistance force of concrete mixture which resulted in controlling the development of initial thermal expansion during temperature-rising stage and restraining shrinkage deformation in constant temperature stage.

3.1.2 Reinforcement effects of nano particles.

N1 and N2 groups' mix proportions are based on M3 group but there are 0.5% NC and 0.5% NS have further replaced of C, respectively, to carry out expansion deformation experiments. Test results are displayed in figure 5 and results indicate that nano particles have significant effects on further restraining the thermal expansion deformation peaks and increasing shrinkage deformations.

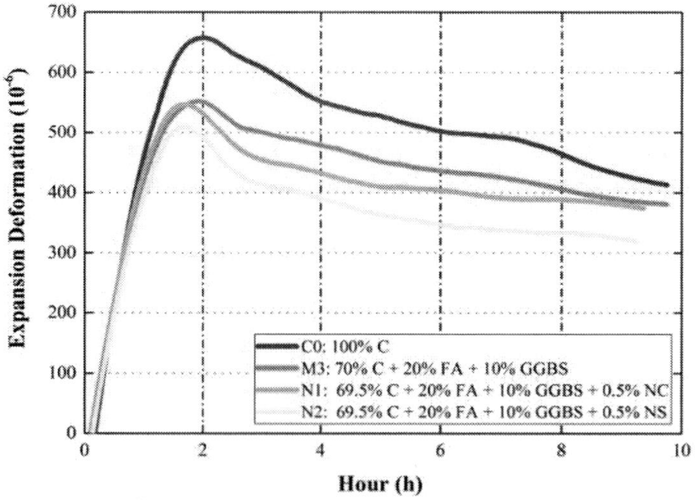

Figure 5. Effects of nano particles on controlling volume deformation.

Figure 5 has shown following changes: when nano particles are blended in concrete mixtures, expansion deformation peaks of N1 and N2 groups decrease in temperature-rising stage, besides, N2 group drops more obviously. Meanwhile, shrinkage deformation of N2 group is more severe while that of N1 group is relatively gentle. When tests are about to end, final expansion value of N1 group nears M3 group but that of N2 group is close to 320μ. Experimental results show the addition of small amount of nano particles help to reduce thermal expansion peaks and promote shrinkage deformations of concrete mixtures during thermal curing process.

Explanations for these experimental results may be illustrated as followed. Nano particles, due to their nano-scale's particle size and huge specific area, efficiently reduce free water content and exert excellent physical filling effect. The initial pore structure of concrete mixture has been improved which provides a stronger resistance force to further restrain the expansion deformation during heating stage. On the other hand, nano particles have not only higher activity but also pozzolanic effect, which promotes cement hydration, changes product phase and fills up more pores. Since hardened cement and early framework of concrete mixture have been formed, the ability to resist thermal expansion deformation has been finally enhanced [11-12]. It is noteworthy that when experiments enter constant temperature stage, shrinkage deformation of N1 group declines fast first and then gradually smoothly because the ability of NC to continuously promote hydration is limited.

3.2 Relationship between initial structural state and deformation

Initial structural state is vital for concrete mixture's thermal deformation characteristics during whole thermal curing process. Initial structural indices are mainly consist of evaporable water content, pore ratio, setting time and compressive strength. In order to expediently carry out experiments, evaporable water content and compressive strength are investigated because these specimens could be synchronously prepared when fabricating volume deformation test specimens. Test results are listed in table 4.

Table 4. Initial structural indices and expansion peaks.

Type	Evaporable Water Content (w% by weight)	Compressive Strength (MPa)	Thermal Expansion Peak (μ)
C0	28.3	2.44	657.0473
M1	28	2.65	614.3168
M2	27.1	2.87	623.8592
M3	25.8	3.11	551.9716
N1	23.2	3.83	546.737
N2	20.5	4.28	509.5949

It can be seen from table 4, concrete mixtures incorporating different cementing components have distinct compressive strengths and evaporable water contents. C0 group owns the highest evaporable water content and lowest compressive strength. When FA and GGBS have replaced cement separately, changing trends of initial indices of M1 and M2 groups are similar. Their evaporable water contents decline and compressive strengths rise. However, GGBS's catalytic effect on early hydration of cement and the place where nucleation happens provided by it also helps to continuously produce more hydration, which results in consuming more free water and heightening bigger early strength of concrete mixture. Because of superimposed compound effect of FA-GGBS, initial pore structure of concrete mixture has been improved and early hydration of cement has been promoted, which further reduces the water content and raise the strength of M3 group to a certain extent [13]. When nano particles are added into concrete, evaporable water contents of N1 and N2 groups significantly decline. Above experimental phenomenon is related to surface properties of NS and NC. These nano particles absorb and consume most free water within concrete mixtures, besides, NS owns a much higher specific area than NC, therefore N2 group has the obviously lowest evaporable water content. With the help of nano particles, superimposed compound effect has been further enhanced and some nano-scale and micro pores would be filled. Then, hydration products of cement and NS would fill up these pores again, which makes concrete specimens become more compact and have a higher compressive strength at the same time.

Besides, according to the data given in table 4, relationships between initial structural indices and thermal expansion peaks have been mathematically fitted and results are displayed in figure 6.

Figure 6 implies that both evaporable water content and compressive strength are exponentially related to thermal expansion peaks. The greater water content, the greater thermal expansion. Besides, the greater initial strength, the smaller thermal expansion. It is obvious that when concrete incorporating mineral admixtures, a reasonable mixing proportion of mineral admixtures is conducive to the generation of superimposed component effect. On this basis, adding nano particles into concrete mixture can further fill internal pores and strengthen a more compact initial structure. On the other hand, with continuing cement hydration and joint effects of mineral admixtures and nano particles, free water is constantly consumed. At the same time, these materials can absorb most of free water which helps to decrease the content of free water in setting stage and evaporable water in curing stage.

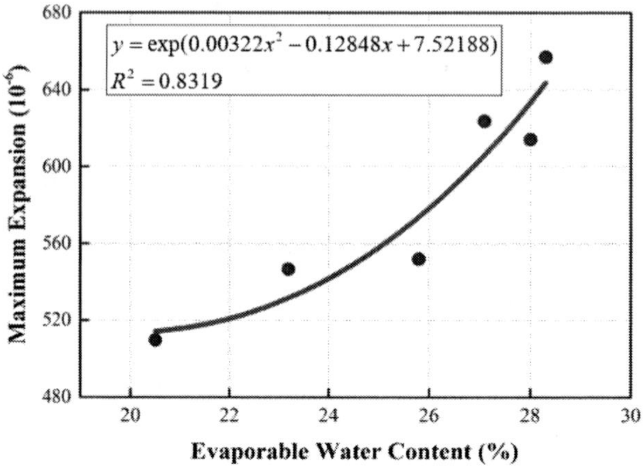

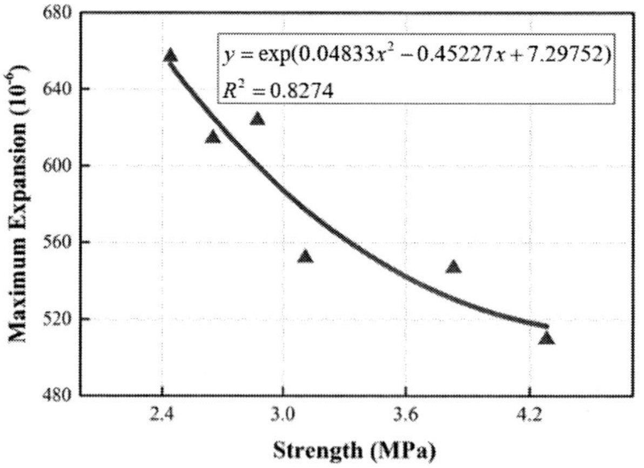

Figure 6. Relationship between initial indices and thermal deformation.

Considering above two aspects, because a denser initial structure has formed and hydration products have been rapidly generated and then fill most nano-scale's and mirco pores, concrete owns a certain strength to resist thermal expansion during thermal curing process. At the same time, consumption of free water greatly reduces vapor evaporation movement which results in further restraining early volume expansion. Therefore, preparing concrete incorporating mineral admixtures and a small amount of nano particles could reduce thermal expansion deformation during the whole thermal curing process.

4 Conclusion

Volume deformation characteristics of concrete mixture could be divided into 2 stages: thermal expansion rising rapidly in temperature-rising stage and shrinkage deformation happens slowly in constant temperature stage. During the whole experimental process, the final feature of concrete specimen is dominated by expansion deformation.

Compared with FA, GGBS has better effects on restraining early thermal expansion and increasing shrinkage. Concrete mixture incorporating 20% FA and 10% GGBS owns small thermal expansion deformation peak and gentle shrinkage. After adding nano particles, the superimposed compound effect has been further enhanced. With the joint effects of mineral admixtures and nano particles, the more free water is absorbed and consumed, the less expansion caused by evaporation would happen. Besides, due to the stronger initial anti-deformation ability and skeleton strength, thermal expansion deformation would be eventually restrained. The recommended dosages of FA, GGBS and NS for replacing cement are 20%, 10% and 0.5%, separately.

Acknowledgments

This study is financially supported by the National Natural Science Foundation of China (Grant No U1534207, 11790283 and 51678568). Authors thank Mr. Q Y Xiao for his great help to accomplish experimental tasks. Meanwhile, authors also appreciate those who are reviewing this paper for your kind concern about construction technology of high-speed railway in China.

References

1. Ma K, Long G and Xie Y 2017 *Case Studies in Construction Materials* **6** 63
2. Long G, Yang J and Xie Y 2017 *Constr. Build. Mater.* **136** 456
3. Ma K, Xie Y and Long G 2014 *Sci. China Technol. Sci.* **57** 1909
4. Long G, Wang M, Xie Y and Ma K 2014 *Sci. China Technol. Sci.* **57** 1902
5. Pater H, Bland C H and Poole A B 1995 *Cem. Concr. Res.* **25** 485
6. Liu B, Luo G and Xie Y 2018 *Constr. Build. Mater.* **167** 359
7. Zou C, Long G, Ma C and Xie Y 2018 *Constr. Build. Mater.* **188** 424
8. Lothenbach B, Winnefeld F, Alder C, Wieland E and Lunk P 2007 *Cem. Concr. Res.* **37** 483
9. Wang Q, Li M and Zhang B 2014 *J. Therm. Anal. Calorim.* **118** 1505
10. He Z, Long G and Xie Y 2012 *J. Cent. South Univ.* **19** 1155
11. Long G, Li Y, Ma C, Xie Y and Shi Ye 2018 *Thermochim. Acta* **664** 108
12. Mei J, Ma B, Tan H, Li H, Liu X, Jiang W, Zhang T and Guo Y 2018 *Constr. Build. Mater.* **171** 83
13. Ho D W S, Chua C W and Tam C T 2003 *Cem. Concr. Res.* **33** 595

Repair Delaminations in Carbon/epoxy Laminates

A.M. Amaro[1], M.A. Neto[1], A.J.R. Loureiro[1], M.P. Silva[2, a] and P.N.B. Reis[2]

[1]*CEMMPRE, Depart. of Mechanical Engineering, University of Coimbra, Rua Luís Reis Santos, Pinhal Marrocos, 3030-788 Coimbra, Portugal*
[2]*C-MAST, Depart. of Electromechanical Engineering, University of Beira Interior, Covilhã, Portugal*

Abstract. Delaminations are the most typical damage in composite laminates, and the industrial community is continually searching new techniques to combine efficiency with cost benefits. Therefore, this work intends to analyse the application of screws on repair of composite laminates containing delaminations. It was found that this technique is beneficial for delaminations with lengths higher than 15 mm. For example, considering delaminations with 20 mm, the compressive load increases around 46.4% relatively to the delaminated samples. However, for small lengths, this technique is not viable due to the stress concentration induced by the hole.

1 Introduction

Delaminations between different oriented layers are the most typical damage that occurs in composite laminates. In this context, literature reports several techniques with the purpose to restore the damaged component for a usable condition.

Traditionally it is used the bonded repairs, where the adhesive layer transfer some load from one part of the component to other one reducing, in this case, the stresses in the repaired component. Mainly two types of patch configurations are available: the scarf patches and the overlap patch [1-4]. However, according to Soutis *et al.* [2], the external patch technique is simpler to apply and is less critical in nature than a scarf method.

The resin injection repair is also regarded as a temporary measure to prevent the spread of damage until further maintenance can be performed [5, 6]. For this purpose, the novel cyanoacrylates bring significant benefits due to the very low viscosity at room temperature and relatively high glass transition temperature [6, 7]. However, some inconvenient are preferred by Wu *et al.* [8] due to the fact that during the injection the surfaces of the delaminations can be contaminated by the environment, making any bonding ineffective. A more effective alternative is the through-thickness reinforcement repair (TTRR), proposed by Li *et al.* [9]. In this case, holes are drilled through the delaminated region and pins bonded into these holes. According to the authors, it is hypothesised that the pins will effectively shorten the length of the sub-laminates and increases the load at which they buckle [9].

Finally, self–healing is a novel alternative to damage tolerant design and removes the need to perform temporary repairs to damaged structures, where an autonomous healing functionality into a composite material is made [10]. In this case the damaged structure is repaired by materials already

[a] Corresponding author: marco.silva@ubi.pt

contained within it, analogous to the biological healing process in living organisms [11]. An ideal self-healing material is capable to restore the material's performance without affect the initial mechanical properties [12].

However, the industrial community is continually open to new techniques, especially when they combine the efficiency with cost benefits. In this context, the aim of the present work is to study the efficiency of delaminations repaired by application of screws. It is known that the delaminations do not affect the laminate behaviour if the adjacent faces tend to close under normal compression. Based in a similar philosophy to the through-thickness reinforcement method, it is expected to restore the mechanical strength required to permit the component operation until a permanent repair can be done.

2 Materials and Procedure

Composite laminates were manufactured using 12 balanced woven bi-directional layers of carbon fibres (196 g/m^2) all of them with the same orientation 0/90°, and an epoxy resin matrix. Details about the manufacture process of the composites laminates can be found in [13]. Delaminations were simulated by a Teflon layer, at the middle of the plate's thickness, with lengths of 2, 5, 10, 15, and 20 mm as described in [14]. Similar samples were manufactured and, using a special drill, open holes with 3 mm of diameter were introduced in the middle of the delamination length. M3 screws were inserted as shown in Figure 1.

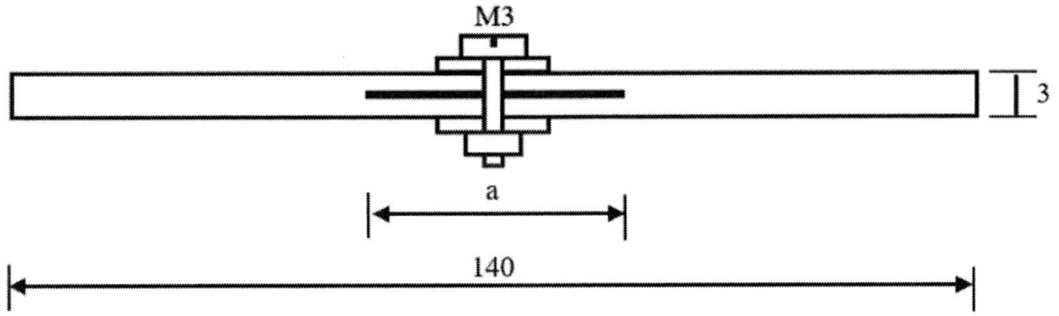

Figure 1. Geometry of the specimens in mm. The samples width is 140 mm and the delamination length (a) of 2, 5, 15, 20 mm.

The compressive tests were performed using an Instron 4206 universal tensile testing machine in accordance with the standard ASTM D 3410-03. All the tests were performed at room temperature with a strain rate of 1.5 mm/min. Global strain data of the specimens was evaluated using a 50 mm gauge length clip-on extensometer in a length between grips of 80 mm. Local strain data acquisition was performed using an Aramis 3D 5M optical system (GOM GmbH) with Digital Image Correlation (DIC). So, full-field displacements of the overlap region could be measured. It should be noted that before testing, the specimens were prepared by applying a random black speckle pattern, over the previously mat white painted surface of the samples, in order to enable data acquisition by DIC [15].

3 Results and discussion

Figure 2 shows typical compressive load-displacement curves relatively to the specimens with a delamination of 15 mm, however, they are representative of all tests. The maximum loads obtained are compared in Figure 3, where symbols represent the average values and the scatter bands the maximum and minimum values obtained.

It is possible to observe a significant decreasing of the maximum load with the presence of delaminations and with their length. For example, between 15 and 20 mm, the compressive strength presents values very close and around 52% lower than the ones observed for the control samples.

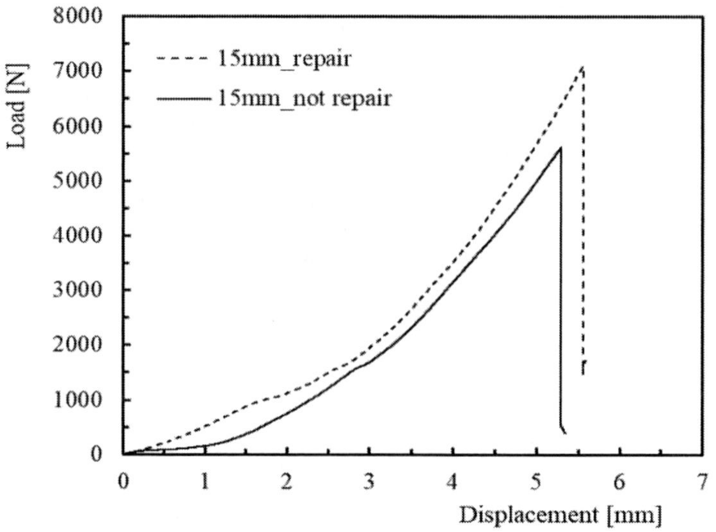

Figure 2. Typical load-displacement curves obtained from the compressive tests.

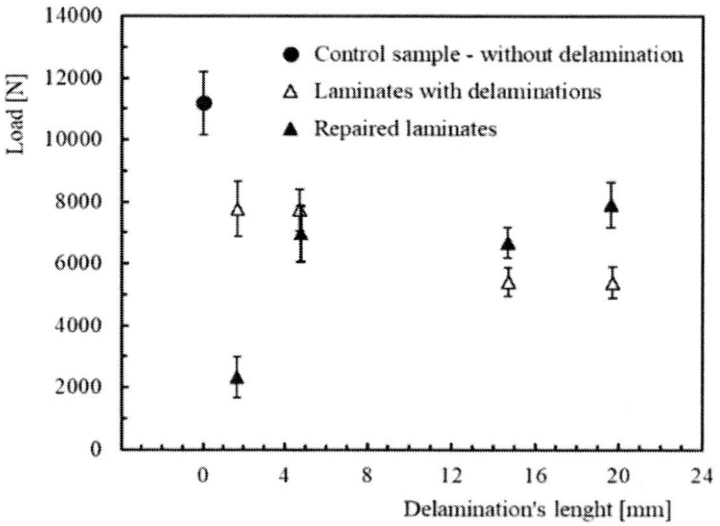

Figure 3. Comparison between laminates with delaminations against repaired laminates in terms of maximum compressive load.

These results are in good agreement with a previous study developed by the authors, where a significant reduction of the compressive strength was observed with the presence of similar defects [16]. Stress concentration promoted by the delaminations is responsible by the decrease observed on the compressive strength [17], however, the failure mechanisms show to be determinant on such process. According with the literature, delaminations under compressive loads can cause buckling in one of three modes: global instability/buckling of the laminate, local instability (buckling of the sub-laminate), or a combination of them [18, 19]. However, the delamination shape [20-22] and the laminate stacking sequence [23] show to be an important factor on the failure mode observed. Amaro et al. [16], for example, observed local buckling as the main failure in laminates containing through-the-width delaminations, i.e., buckling of the sub-laminate delaminated. In this case, delaminations tend to grow rapidly leading to the structural failure [24]. Similar phenomenon was found in the present study, as shown in Figure 4. Figure 4a) shows the typical final failure occurred in samples with delaminations, while Fig 4b) shows the delamination with 15 mm of length, respective propagation during the initial loading values and, simultaneously, the initial buckling of a sub laminate delaminated.

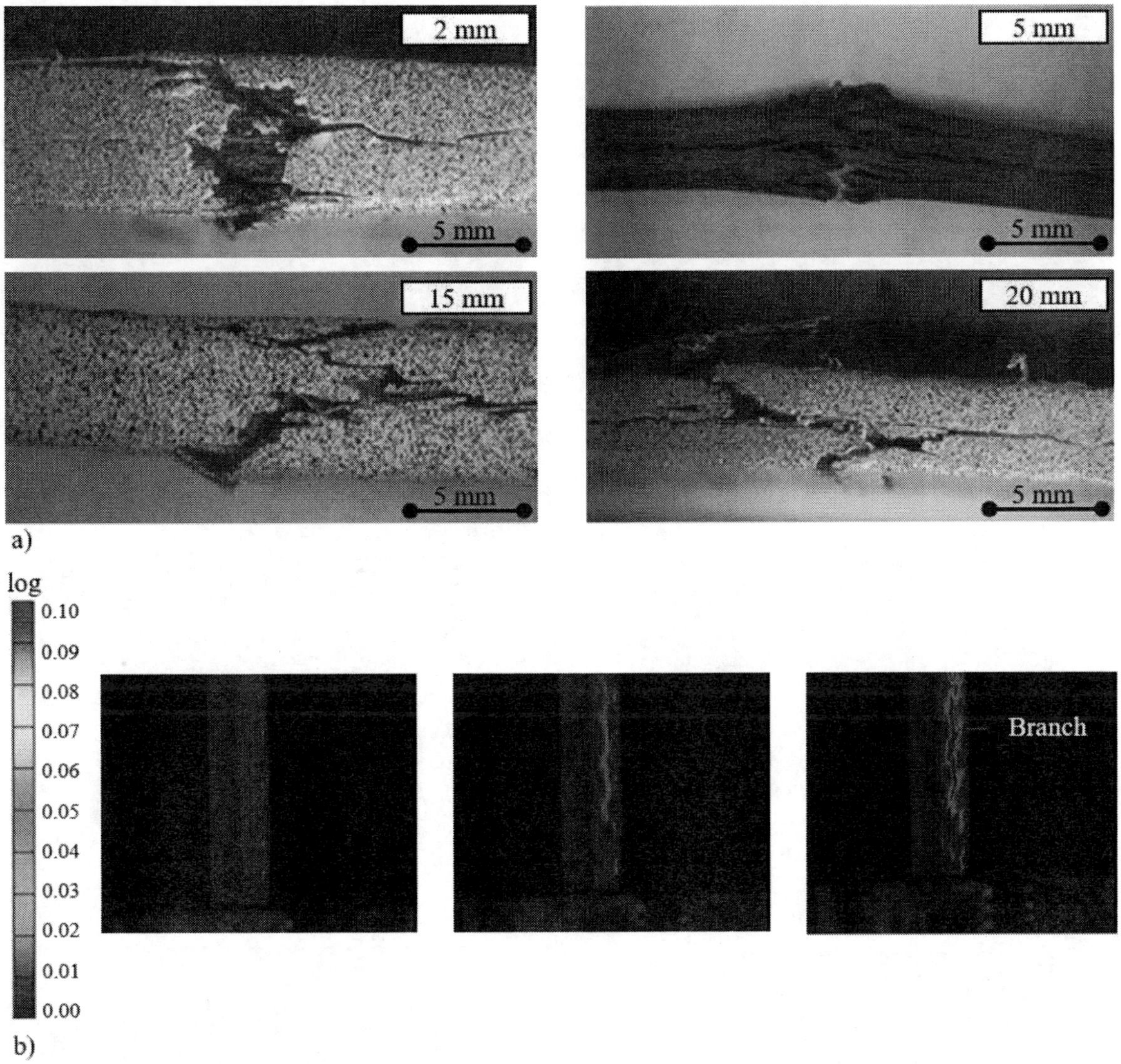

Figure 4. a) Typical final failures for laminates with delamination; b) Sequence of the damage progretion for the initial loading obtained for 15 mm of delaminations's length.

Relatively to the samples repaired by screws, as shown in Figure 3, it is possible to observe a significant drop of the compressive strength for samples with delaminations of 2 mm. A decreasing around 70% is observed comparatively to the samples with delamination, which is expected as consequence of the hole necessary to the screw to be higher than the delamination's length. Similar tendency was observed for delaminations with 5 mm but, in this case, only a slight decreasing about 4.5% was found. However, for higher values a positive effect of the screw is visible, indicating this technique as solution to repair laminates composites with delaminations. For delaminations with 15 and 20 mm of length, the compressive load increases around 23.4% and 46.4%, respectively, compared to the delaminated samples.

Figure 5 shows the typical final failures observed for samples repaired, which are different from the ones occurred for samples with delaminations.

a)

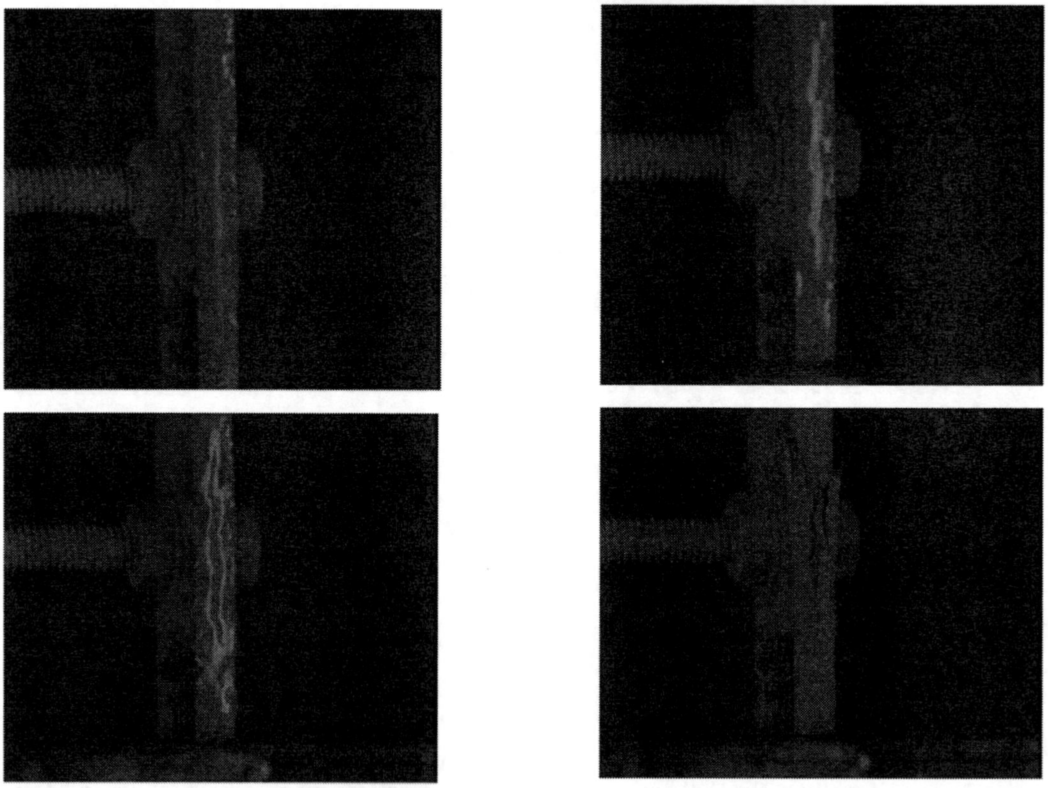

b)

Figure 5. a) Typical final failures for repaired laminates; b) Sequence of the damage progretion obtained for 15 mm of delaminations's length.

In this case, the screw prevents the opening of the delamination and the buckling is not the main damage mechanism. As shown in Figure 5a), the damage is essentially localized in the region of the screw, for all configurations, because the holes promote stress concentrations responsible by the damages occurred around their border [25]. Figure 5b) evidences the evolution of the damage for samples with delaminations of 15 mm, where is visible the severity of the damage near the screw.

4 Conclusions

The focus of this study was to evaluate the possibility to repair composite laminates containing delaminations. For this purpose, a new technique using screws was suggested in order to solve the most typical damage in such materials.

It was possible to conclude that the present technique is not viable for small delamination lengths, as consequence of the stress concentration induced by the hole, but for delaminations higher than 15 mm shows to be efficient. However, more research works should be performed to obtain a complete knowledge, in order to make feasible its application on real conditions.

References

1. P. Cheng, X.-J. Gong, D. Hearn, S. Aivazzadeh, Compos. Struct. **93**, 582 (2011).
2. C. Soutis, D.-M. Duan, P. Goutas, Compos. Struct. **45**, 289 (1999).
3. F.Z. Hu, C. Soutis, Compos Sci Technol. **60**, 1103 (2000).
4. S.R.M. Coelho, P.N.B. Reis, J.A.M. Ferreira, A.M. Pereira, Compos. Struct. **168**, 259 (2017).
5. M. Hautier, D. Lévêque, C. Huchette, P. Olivier, Plast. Rubber Compos. 39, 200 (2010).
6. S.M. Bleay, C.B. Loader, V.J. Hawyes, L. Humberstone, P.T. Curtis, Compos Part A-Appl S. **32**, 1767 (2001).
7. P.G. Slattery, C.T. McCarthy, R.M. O'Higgins, Compos. Struct. **153**, 1 (2016).
8. K.-W. Wu, C.-L. Lee, Y.-C. Chang, C.-L. Ong, Mater. Chem. Phys. **43**, 173 (1996).
9. H.C.H. Li, P.J. Callus, I. Herszberg, Compos. Struct. **75**, 539 (2006).
10. G. Williams, R. Trask, I. Bond, Compos Part A-Appl S. **38**, 1525 (2007).
11. J.W.C. Pang, I.P. Bond, Compos. Sci. Technol. **65**, 1791 (2005).
12. D.Y. Wu, S. Meure, D. Solomon, Prog. Polym. Sci. **33**, 479 (2008).
13. P.N.B. Reis, J.A.M. Ferreira, J.D.M. Costa, M.O.W. Richardson, Compos. Sci. Technol. **69**, 1541 (2009).
14. P.N.B. Reis, , J.A.M. Ferreira, F.V. Antunes, M.O.W. Richardson, J. Compos. Mater. **43**, 2609 (2009).
15. C. Leitão, M.I. Costa, K. Khanijomdi, D.M. Rodrigues, Mater. Des. **51**, 968 (2013).
16. A.M. Amaro, P.N.B. Reis, M.F.S.F. de Moura, M.A. Neto, Fiber Polym. **15**, 560 (2014).
17. T.K. O'Brien, M. Rigamonti, C. Zanotti, Int. J. Fatigue **11**, 379 (1989).
18. C. Soutis, P.T. Curtis, Compos. Sci. Thecnol. **56**, 677 (1996).
19. C. Soutis, F.C. Smith, F.L. Matthews, Compos. Part A-Appl S. **31**, 531 (2000).
20. M.J. Pavier, W.T. Chester, Composites **21**, 23 (1990).
21. E.S. Greenhalgh, Compos. Struct. **23**, 165 (1993).
22. G.J. Short, F.J. Guild, M.J. Pavier, Compos. Sci. Technol. **61**, 2075 (2001).
23. A. Tafreshi, T. Oswald, Int. J. Pres. Ves. Pip. **80**, 9 (2003).
24. S.-Y. Lee, D.-Y. Park, Int. J. Solids Struct **44**, 8006 (2007).
25. C. Soutis, N.A. Fleck, P.A. Smith, J. Compos. Mater. **25**, 1476 (1991).

Models of Geometrically "Stabilized" Laser Cavity

A. Aissani[1,a], S. Leghmizi[1] and K. Battou[2]

[1] *Laboratory of Quantum Electronics, Faculty of Physics-USTHB, B.P.32,El Alia, 16111, Bab Ezzouar, Algiers, Algeria.*
[2] *CSViamonde, 116 Cornelius Parkway, Toronto M6L 2K5, Canada*

Abstract: The Lamb-dip characterizing the line shape emitted by a gas laser is often used in spectroscopy of materials as a reference for frequency measurements. For such lasers, the frequency control is performed on the Lamb-dip. It is therefore essential, for accurate measurements, that its frequency matches with the laser resonance frequency. This is only possible if the emitted line shape is symmetrical, which is not usually the case. Indeed, the lens effects induced in the laser amplifying medium, which are due to the population and the saturation inhomogeneities, generally produce an asymmetrical emitted line shape. So, the frequency of the Lamb-dip is shifted compared to the central frequency.

In this work, we will first revisit the model given in the literature, in order to highlight the limit of its validity, and then we will propose through an appropriate choice of the cavity geometry, a "stabilized" cavity model giving rise to a symmetrical line shape even when the control parameters vary.

1 Introduction

In several areas of physics [1-4], the Lamb dip characterizing the line shape of a gas laser is used as a reference for frequency and length measurements. These measurements would be more accurate if the line shape was symmetrical, in which case the center frequency of the laser would coincide with that of the Lamb dip [5]. In fact, several causes are at the origin of this asymmetry [6-9]. These combine to create population and saturation inhomogeneities, which induce lens effects that modify the wave mode diameter. The variations as a function of the frequency of these parameters are then at the origin of the asymmetry of the line [10]. However, geometrical parameters [1, 11] can modulate the competition between these two types of antagonistic heterogeneities and favor one or the other. It is for this purpose that several models of cavities, called "stabilized", have been proposed [12-13].

In this work, we will first revisit one of these two models [12] and then, based on the perturbed Gaussian beam theory [14] and by making a judicious choice of geometric parameters, we will propose a geometric cavity structure giving a symmetrical output line shape, even when control parameters vary.

2 Theoretical Model

The theoretical calculation described below concerns the laser cavity shown schematically in Fig.1. It consists of a plane mirror M_1 and a curved mirror M_2 of curvature radius R, separated by a distance D

[a] Corresponding author : ah.aissani@yahoo.com

and placed respectively at z_1 and z_2, an amplifier tube of length L_g, placed between z_3 and z_4, and a diaphragm of diameter $2b$ placed on the side of the output mirror M₂. The mirrors reflective powers are r_A and r_B, respectively. In order to have a fixed linear polarization, the ampifier tube is closed by two Brewster angle windows.

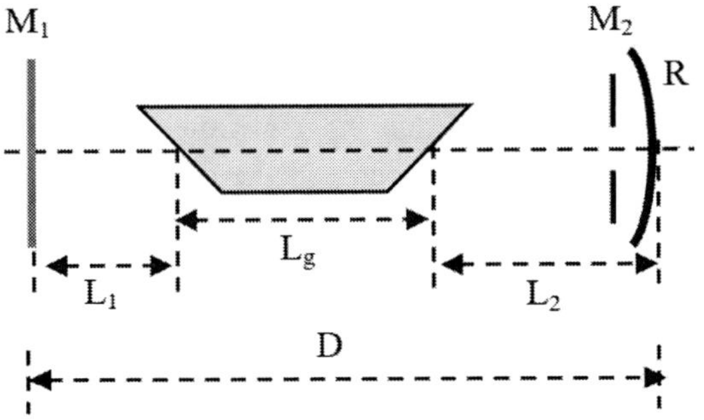

Figure 1: Schematic representation of the cavity

The amplitude of the resonant field is obtained from the propagation equation. Its expression, for the forward wave, can be written in the form:

$$E(\rho,\zeta) = \frac{E_0}{1+2i\zeta} exp\left[i\left(k\frac{\omega}{c}W_0^2\right)\zeta\right] exp\left[\frac{-\rho^2}{1+2i\zeta}\right] exp\left[-i\left(p - \frac{\rho^2}{2}q\right)\right] \quad (1)$$

Where: E_0 is a constant, k the propagation constant, p and q represent, respectively, the perturbations of the longitudinal and transversal parameters of the wave, $2W_0$ is the mode diameter at the beam waist, and (ρ,ζ) are the reduced coordinates connected to the cylindrical coordinates (r,z) by:

$$\rho = \frac{r}{W_0};\ \zeta = \frac{cz}{\omega W_0^2} \quad (2)$$

The proposed cavity is composed of three regions. In region 1: $z_1 \le z \le z_3$, in region 2: $z_3 \le z \le z_4$ and in region 3, $z_4 \le z \le z_2$. The propagation constant k, it is equal to $\left(\frac{\omega}{c}\right)$ in the empty regions 1 and 3 and k_g in the region 2 containing the active medium. In addition, the oscillating field in the cavity is composed of a forward and a backward waves, respectively indexed "f" and "b" in the next equation. The equation (1) takes the explicit form below:

$$^{\mathrm{m}}E_{f,b}(\rho,\zeta) = \left(\frac{^{\mathrm{m}}E_{0f,b}}{1\pm 2i\zeta}\right) exp\left[\pm i\left(k\frac{\omega}{c}W_0^2\right)\zeta\right] exp\left[\frac{-\rho^2}{1\pm 2i\zeta}\right] exp\left[-i\left(^{\mathrm{m}}p_{f,b} \mp \frac{\rho^2}{2}{}^{\mathrm{m}}q_{f,b}\right)\right] \quad (3)$$

The upper sign in the symbols "$\pm$" or "$\mp$" corresponds to the forward wave and the lower one to the backward wave.

This expression allows us to easily determine the field at the output of the cavity. We will assume that the transmission coefficients at the boundary of two neighboring regions are equal to 1. In addition, the continuity relations of the field make it possible to write:

$$E_{f,b}(0,\zeta_j) = {}^{\mathrm{m}}E_{f,b}(0,\zeta_j) \quad (4)$$

Where ζ_j is the reduced coordinate at the border of two neighboring regions (j = 3 or 4) and m is the region number.

Given these considerations, we obtain at ζ_2:

$$^{3}E_f(\rho,\zeta_2) = \left[\frac{1+2i\zeta_1}{1+2i\zeta_2}\right] {}^{1}E_f(0,\zeta_1) exp\left[i\left(\frac{\omega}{c}W_0\right)^2 (l_1 + l_2)\zeta + i\frac{\omega}{c}W_0^2 k_g l_g\right] exp(-ip_F)$$
$$.\, exp\left[\frac{-\rho^2}{1+2i\zeta}\right] exp\left(-i\frac{^{\mathrm{m}}q_{f,b}}{2}\rho^2\right) \quad (5)$$

Where $p_F = ({}^1p_{f_3} - {}^1p_{f_1}) + ({}^2p_{f_4} - {}^2p_{f_3}) + ({}^3p_{f_2} - {}^3p_{f_4})$ and l_1, l_2, l_g are the reduced lengths corresponding respectively to L_1, L_2 and L_g.

The corresponding intensity which is proportional to the square of the modulus of the electric field amplitude is defined as:

$$I_f(\rho,\zeta_2) = \left[\frac{1+4\zeta_1^2}{1+4\zeta_2^2}\right] I_f(0,\zeta_1)\, exp\left[-2\left(\frac{\omega}{c}W_0^2\right)k_g^i l_g + 2p_F^i\right] exp\left[-\frac{2}{1+4\zeta_2^2}\rho^2\right] exp(-{}^3q_{f_2}^i\rho^2) \quad (6)$$

Where k_g^i, p_F^i, and $q_{f_2}^i$ designate the imaginary parts of these different quantities.

Since the asymmetry of the line shape depends on the position of the detector at the output of the laser [15], we will be interested in the emitted total power $P_f(\zeta_2)$. If b is the radius of the diaphragm (the corresponding reduced length is then $\left(\frac{b}{W_0}\right)$, we have :

$$P_f(\zeta_2) = W_0^2 \int_0^{2\pi} d\varphi \int_0^{\frac{b}{W_0}} I_f(\rho,\zeta_2)\rho d\rho \quad (7)$$

After some analytical calculations, we obtain:

$$P_f(\zeta_2) = \pi W_0^2 I_f(0,\zeta_1)\left[\frac{(1+4\zeta_2^2)}{2+(1+4\zeta_2^2)\,{}^3q_{f_2}^i}\right] exp\left[-2\left(\frac{\omega}{c}W_0^2\right)k_g^i l_g + 2p_F^i\right]$$

$$\left[1 - exp\left[-\frac{b^2}{W_0^2}\frac{2+(1+4\zeta_2^2)\,{}^3q_{f_2}^i}{(1+4\zeta_2^2)}\right]\right] \quad (8)$$

This equation shows that the total power $P_f(\zeta_2)$ at the side 2 of the laser depends upon the total power $\left(\pi W_0^2 I_f(0,\zeta_1)\right)$ at the side 1, modified by the transmission factor across the diaphragm (last term in square brackets), the gain by the exponential containing k_g^i and the saturation term $2p_F^i$, and the geometric parameters of the cavity.

The obtained expression contains the longitudinal and transverse perturbations of the Gaussian beam that can be obtained using their evolution equations and continuity conditions at the boundary of two regions. We will use the equation 8, in our numerical calculations to plot the line shapes at the output of the laser.

3 Numerical Results

The numerical calculation concerns a single mode He-Ne laser operating at 3.39 μm and it's based on the perturbed Gaussian beam model [14]. The cavity (see Fig.1) is hemispherical, of length D. The reflection coefficients are $r_A = 1$ for the plane mirror and $r_B = 0.8$ for the concave one. The curvature radius for the output mirror M_2 is equal to $R = 600\ mm$. The amplifier tube, of length L_g, is closed by two windows at Brewster angle. In order to select the Gaussian fundamental mode TEM_{00}, the diameter $2b$ of the diaphragm was chosen to be equal to about 1.2 times the mode diameter at the laser output [12].

In a first step, we wanted to verify the model of stabilized cavity proposed by the authors of the reference [12]. In this model, the active medium is assumed to a thin lens centered in the cavity. So, the approach followed is to assume that the amplifier tube is centered in the cavity to find a value G_0 of the gain which gives a symmetrical line shape and, if so, to test other gain values, below and above G_0, to see if the symmetry of the line shape is preserved. The length D of the cavity is fixed only by the value R of the curvature radius of the curved mirror through the relation $D = \frac{4}{5}R$ [12], which corresponds in our case to $D = 480\ mm$. In the following the length of the active tube is fixed to $L_g = 200\ mm$.

The Fig.2 shows, for three appropriately chosen gain values, the total power at the laser output as a function of the reduced frequency $x = (\omega - \omega_0)/\Gamma$, where ω_0 is the laser center frequency and Γ is the Doppler half-width. We see that the line shape is symmetric for $G_0 = 1.86$, as predicted by the authors. On the other hand, for a smaller gain $G = 1.80$, the saturation inhomogeneity outweighs the population one. The asymmetry of line shape has a maximum on the low frequencies side (LF asymmetry type), while, for a higher gain $G = 2.10$, population effects outweigh the saturation one and the maximum is on the high frequencies side (HF asymmetry type). This shows clearly the switch of the asymmetry from one type to the other when the gain varies. We conclude that for G_0, population and saturation effects compensate each other, giving a symmetrical line shape. But this symmetry disappears as soon as the gain G changes. It is therefore obvious that the proposed model does not allow to obtain a so called "stabilized cavity".

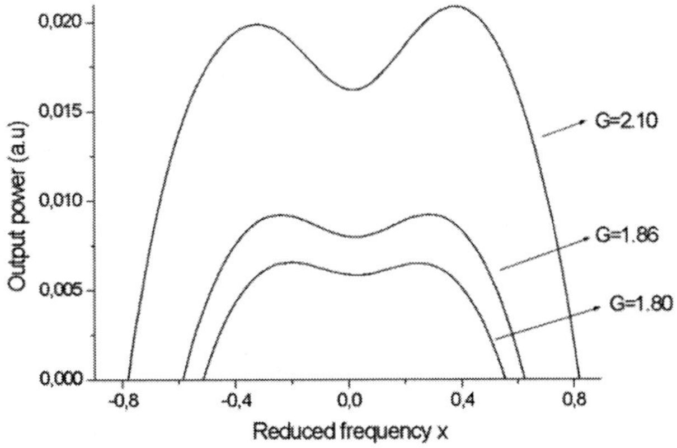

Figure 2: The line shape of the laser output versus the reduced frequency x

To achieve the main part of this paper, we will now focus on the study of the case where the gain varies, and we will clearly show that this variation does not affect the asymmetry, in contrary to the previous model. After several simulations, the cavity length is set to $D = 500\ mm$ and contains an active tube of length $L_g = 150\ mm$. The rest of the settings are unchanged. The gain G has been set to 2.1. This choice is motivated by the fact that this value gives rise to a switch of the asymmetry when the position of the active tube varies and gives line shapes with a pronounced Lamb-dip.

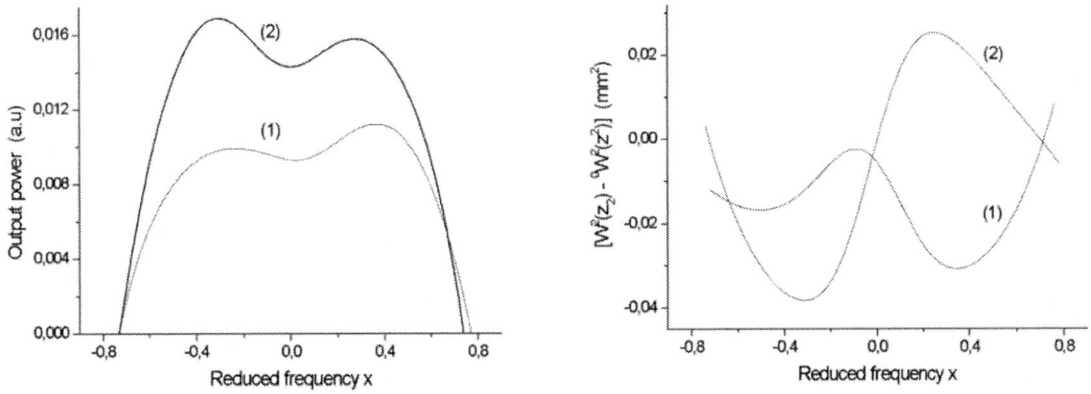

(a) (b)

Figure 3: (a) Line shape and (b) variations of the quantity $[W^2(z_2)-{}^0W^2(z^2)]$ versus the reduced frequency for two positions of tube: (1) near the plane mirror, (2) near the concave mirror.

Fig.3a and 3b show respectively the line shapes obtained for two positions of the tube (plane side and curved side), and the variations of the corresponding mode diameter. As it can be seen, the line shape, initially of HF type becomes LF type when the tube is brought from the plane mirror to the concave mirror. The Garside effect plays, in this case, its entire role as shown in Fig.3b. The variation of asymmetry obtained implies the existence of an optimal position for which the line shape becomes symmetrical. In the following we will characterize the rate of the asymmetry by η which is calculated according to $\eta = 2(I_{HF} - I_{LF})/(I_{HF} + I_{LF})$, where I_{HF} and I_{LF} are the maximums of power at the high and low frequencies, respectively. The variations of η as a function of the position z_3 of the tube allows the determination of the position $z_{3.opt}$ which gives $\eta = 0$ corresponding to a symmetrical line shape. This gives $z_{3.opt} = 174\ mm$. However, it must be verified that, for this optimal position, the line shape remains symmetrical, even when the value G of the gain is changed. Fig.4a illustrates this situation. Unlike the previous model, the line shape keeps its symmetry, even for high gains. This model shows clearly that we don't have a competition between the population and saturation effects, but rather a total cancellation of the variations of the mode diameter. This is confirmed through Fig.4b where the line shapes are perfectly symmetrical with respect to the center frequency. The Garside effect therefore acts in the same way for the low and high frequencies and the line shape keeps its symmetry. We can now say that, for this particular choice of cavity geometry, the effects of population and saturation inhomogeneities which are at the origin of the line shape asymmetry, do not compensate each other as it was the case in the model presented in reference [12] but are canceled simultaneously. The cavity is thus stabilized with respect to its transverse parameter and, in this case, the frequency of the Lamb-dip center matches exactly with the laser resonant frequency. This may be useful for the use of such a cavity as a standard of length and frequency and can be used in spectroscopy of materials as a reference for frequency measurements.

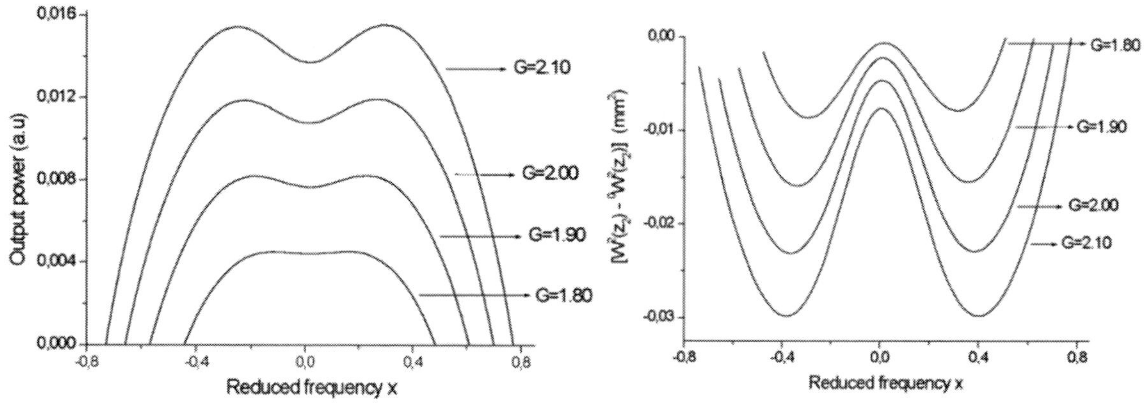

Figure 4: (a) Line shape and (b) variations of the quantity $[W^2(z_2)-{}^0W^2(z_2)]$ versus the reduced frequency for several gain values.

4 Conclusion

This study has highlighted the influence of several interdependent parameters on the line asymmetry. This complexity limits the use of simple theoretical models obtained from analytical processing. The saturation and population inhomogeneities are at the origin of the asymmetry of the line shape. However, the geometrical parameters of the cavity and the gain influence these two types of inhomogeneities and compete to favor one or the other asymmetry type. In some favorable cases, the two effects compensate each other, for a given gain, and the line shape is symmetrical. This corresponds to a simple compensation of antagonistic effects Unfortunately, the symmetry disappears as soon as the gain varies

Through the model we developed in the present work, we have shown, that a judicious choice of the geometrical parameters of the cavity, allows to cancel simultaneously the effects due to the population and the saturation inhomogeneities. The approach adopted allowed us to obtain a geometric configuration giving rise to a symmetrical line shape at the output of the cavity, and whatever the variation of the value of the gain. The obtained cavity is therefore well and truly stabilized.

References

1. A. Aissani, B. Meziane, K. Battou and O. Ziane, "*Spectral-line asymmetry in double-output-cavity lasers*", *Optics & Laser Technology*, vol. 42, no. 2, pp. 439-446, Oct 2009.
2. Z.D. Sun, R.M Lees and L.H. Xu, "*Precision Lamb-dip infrared spectra of the C-N stretching band of CH_3NH_2 with a CO_2-laser microwave-sideband spectrometer*". *J. Chem. Phys.*, vol. 132, 1-6, May 2010.
3. Z.D Sun, and al., "*Lamb-dip spectroscopy of the C−N stretching band of methylamine by using frequency-tunable microwave sidebands of CO_2 laser lines", Sci. Rep.*,vol. 6, 34270; doi: 10.1038/srep34270, 2016.
4. S. Welzel, F. Hempel, M. Hübner ,N. Lang, P. B. Davies and J. Röpcke, "*Quantum Cascade Laser Absorption Spectroscopy as a Plasma Diagnostic Tool: An Overview*",*Sensors* ,vol 10, no.7, pp.6861-6900, 2010.
5. S. L. Ma, S. Picard, M. Zucco, J.M. Chartier, L. Robertsson, R.S. Windeler," *Direct measurement of the absolute frequency of the international reference laser BIPM4*" *Metrologia*, vol. 41, pp. 65-8, 2004.
6. A. Szoke, A. Javan, "*Effects of collisions on saturation behaviour of the 1.15 um transition of Ne studied with He-Ue laser*" *Phys Rev*, vol.145, no.1, pp. 137-147, May1966.
7. R.H. Cordover, P.A. Bouczyk, *"Effects of Collisions on the Saturation Behavior of the 6328 A° Transition of Ne Studied with a He-Ne Laser"*, *Phys Rev*, vol. 188, no.2, pp. 696-700, Dec 1969.
8. A.Szoke, A. Javan, " *Isotope Shift and Saturation Behavior of the 1.15-μ Transition of Ne", Phys. Rev. Lett.* vol. 10, no. 12, pp. 521-524, Jun. 1963.
9. A. V. Miranov, " *Observation of a Symmetrical Lamb Dip for a He−Ne Laser with a Natural Mixture of Neon Isotopes" Optics and Spectroscopy*, vol. 92, pp. 473-474, no 3, March 2002.
10. G.Stephan, H. Taleb, F. Legros, C. Pesty, *" Mise en évidence expérimentale et théorique de deux causes de dissymétrie dans la raie d'émission d'un laser à gaz" J. Phys.* vol. 43, pp. 255 – 261, 1982.
11. G. Stephan, H. Trumper "*Gaussian beam parameters and their stabilization in gas lasers", Phys. Rev.* A, vol. 28, no.6, pp.3450-3458, Dec 1983.
12. A. Le Floch, J.M. Lenormand, R. LE Naour andJ. P. Tache, » *a critical geometry for lasers with internal lenslike Effects*" *J. Phys. Lett.*vol. 43, no 14, pp. 493-498, 1982.
13. P. Cerez, R. Felder, " *Gas-lens effect and cavity design of some frequency-stabilized He–Ne lasers", Appl. Opt.* vol. 22, pp. 1251-1256, 1983.
14. G. Stephan, H. Trumper, *"Inhomogeneity Effects in a Gas Laser", Phys. Rev.* vol. A28, no.4, pp.2344-2362, Oct. 1983.
15. A. Kellou, G. Stephan, "*Etude du champ proche d'un laser diaphragmé",* Applied Optics Vol. 26 (1987) 76-90.

Experimental Evaluation and Numerical Modelling Residual Stresses in Glass Panel

Jüri Majak[1,a], Johan Anton[3,4], Erko Õunapuu[1], Fabio Auriemma[1], Meelis Pohlak[1], Martin Eerme[1] and Aleksander Klauson[2]

[1]*Department of Mechanical and Industrial Engineering, Tallinn University of Technology, 19086, Tallinn, Estonia*
[2]*Department of Civil Engineering and Architecture, Tallinn University of Technology, 19086, Tallinn, Estonia*
[3]*Department of Cybernetics, Tallinn University of Technology, 19086, Tallinn, Estonia*
[4]*GlasStress Ltd, 12618, Tallinn, Estonia*

Abstract. During last decade increased usage of laminated composite glass structures, also annealed and tempered glass can be observed in civil engineering, automobile and space structures, solar panels, etc. Latter trend is caused by high strength properties of laminated glass, also sound and vibration attenuation capabilities. However, heat treatment of glass causes residual stresses, which are not often covered in structural analysis. Current study is focused on experimental evaluation and numerical modelling of residual stresses in glass panels.

1 Introduction

The non-destructive testing methods are preferred as resource saving and sustainable. The PVB interlayers have key role in aging of LGCP laminates and need special attention since in a number of application long term exploitation of LGCP panels has been foreseen (solar panels, windows). The elastic material properties of the PVB interlayer determined experimentally are used as simulation input parameters in order to predict vibration attenuation capability at concurrent frequency excitation and displacement.

The main advantage of the laminated glass composite panel, over float glass is its high strength properties, especially fact that it does not break into small pieces even in the case of impact loading cases [1]. The residual stresses arise in glass during cooling down from the crystallization temperature. The residual stresses are caused by the thermal expansion and the elastic mismatch between the crystalline and glassy phases [2,3]. The strengthening effect is achieved with accurate cooling the glass after heat treatment. The fast cooling causes outer layers of the glass to compress and close micro cracks in the surface of the glass. The heat treated glass is used in applications with safety glass requirements [4]. The heat treated glass, especially laminated glass composite panels (LGCP) are used in structures where glass is the main load bearing element. The LGCP consists of glass layers and at least one flexible interlayer. The interlayer provides higher post failure safety of LGCP due to ability to hold broken-off pieces of glass together [5]. Multifunctional LGCP with improved strength and acoustic performance can be obtained by using novel acoustic interlayers covering strengthening and

[a] Corresponding author : juri.majak@taltech.ee

sound attenuation properties simultaneously. Development of new sound absorptive element for wide class of structures and applications is emerging research area [6-9].

In the current study an attempt is made to combine the know-how and capabilities of the private company GlasStress Ltd in area of experimental study of residual stresses in glass structures [10-12] and optimization workgroup of TTU (Tallinn University of Technology) in area of structural analysis and design optimization [13-17].

2 Experimental study

The experimental study performed cover evaluation of residual stresses in glass, also measuring the properties of the constituents of glass laminate composite panel for use in further numerical analysis.

2.1 Residual stresses in glass

2.1.1 Measuring residual stresses

Measuring residual stresses has been conducted in private company GlasStress Ltd according to proposed design of experiment. The SCALP-05 and scattered light method were used to determine through-the-thickness stress distribution of annealed and heat treated sodalime glass panels. Three repeatable measurements were done in directions of the long and short sides of the panel at 10 different locations panel. The residual stresses were measured on both sides up to middle surface of the glass panel. The tin side of the glass was detected by employing Tin Side Detector UV1301. The samples tested were 500x200 mm with thicknesses: 4mm, 8 mm and 4+4 mm laminated glass with 0,38 mm PVB interlayer. Through-thickness distribution of the stress component σ_1 is given in Figure 1.

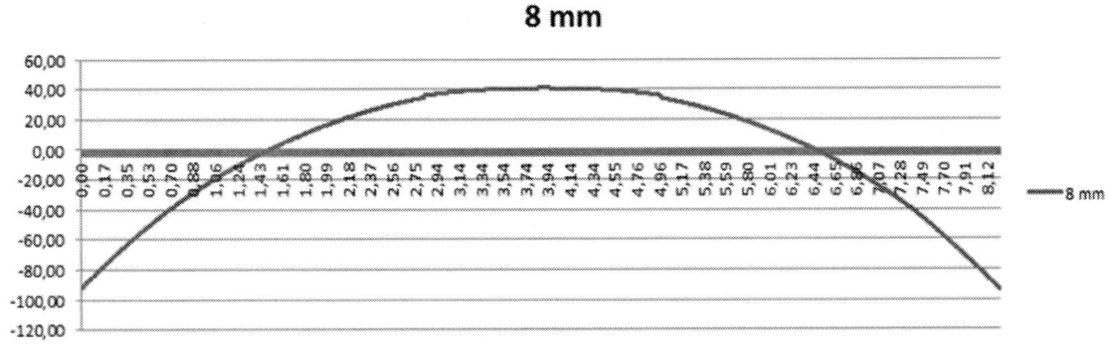

Figure 1. Normal stress component σ_1 - through-thickness distribution

Similarly, symmetric through-thickness distribution of the normal stress component σ_2 has been observed. The experimental data obtained, provide valuable information on behaviour of residual stresses in glass panel. However, these data are not enough for numerical modelling of residual stresses in glass panel due to presence of normal stress components only. Solution of the latter problem is discussed in the following section.

2.1.2 Determining shear stresses in laminated glass

In order to evaluate the shear stresses σ_{12} in glass panel the procedure based on measurement of normal stresses in different rotation angles and application of the plane stress tensor rotation formulas is introduced. The procedure is given as follows:

- The initial coordinate system was specified by long and short sides of the glass specimen;
- The stress components $\sigma_1^{exp\_0}$ and $\sigma_2^{exp\_0}$ were measured in coordinate system rotated 0°;

- The stress components $\sigma_1^{exp\_\theta}$ and $\sigma_2^{exp\_\theta}$ were measured in coordinate system rotated $\theta°$;
- Based on values of the stress components $\sigma_1^{exp\_0}$ and $\sigma_2^{exp\_0}$ the theoretical expected values of the stress components in coordinate system rotated $\theta°$ were computed by employing stress tensor rotation formulas as

$$\sigma_1^{theoret\_\theta} = \sigma_1^{exp\_0} cos^2(\theta) + \sigma_2^{exp\_0} sin^2(\theta) + 2\sigma_{12}^0 \sin(\theta)\cos(\theta), \tag{1}$$

$$\sigma_2^{theoret\_\theta} = \sigma_1^{exp\_0} sin^2(\theta) + \sigma_2^{exp\_0} cos^2(\theta) - 2\sigma_{12}^0 \sin(\theta)\cos(\theta), \tag{2}$$

- Note that the formulas (1)-(2) are not directly applicable, since the value of the shear stress σ_{12}^0 is unknown. For this reason the shear stress σ_{12}^0 was considered as design variable and the differences between experimentally measured normal stresses $\sigma_1^{exp\_\theta}$, $\sigma_2^{exp\_\theta}$ and theoretically computed normal stresses $\sigma_1^{theoret\_\theta}$, $\sigma_2^{theoret\_\theta}$ were subjected to minimization

$$f = (\sigma_1^{theoret\_\theta} - \sigma_1^{exp\_\theta})^2 + (\sigma_2^{theoret\_\theta} - \sigma_2^{exp\_\theta})^2 \rightarrow min. \tag{3}$$

- Applying the necessary optimality conditions for posed optimization problem $\frac{\partial f}{\partial \sigma_{12}^0} = 0$, the value of the shear stress σ_{12}^0 in initial coordinate system was determined as

$$\sigma_{12}^0 = \frac{\left(sin^2(\theta) - cos^2(\theta)\right)\left(\sigma_1^{exp\_0} - \sigma_2^{exp\_0}\right) + \left(\sigma_1^{exp\_\theta} - \sigma_2^{exp\_\theta}\right)}{4\sin(\theta)\cos(\theta)}. \tag{4}$$

Note, that the proposed approach is based on measuring residual stresses in two different angles of the coordinate system and is just one possibility for estimating the value of the shear stress σ_{12}^0. Obviously, the result depends on selection the value of the rotation angle θ. The theoretical approach introduced can be simply extended for a number of rotations of the stress tensor by angles θ_1 to θ_n. However, in latter case the cost needed to pay for higher accuracy is higher number of experiments (the normal stresses should be measured for each value of θ_i).

2.1.3 Error estimate

Having the measured values of the normal stresses $\sigma_1^{exp\_0}$, $\sigma_2^{exp\_0}$ in initial coordinate system and corresponding estimate on shear stress σ_{12}^0 one can apply the stress tensor rotation formulas (1)-(2) for computing the theoretical values of normal stresses $\sigma_1^{theoret\_\theta}$ and $\sigma_2^{theoret\_\theta}$ in new, rotated coordinate system. The errors e_1 and e_2 of the measured normal stresses σ_1 and σ_2 can be estimated as

$$e_1 = \sigma_1^{exp\_\theta} - \sigma_1^{theoret\_\theta}, \tag{5}$$

$$e_2 = \sigma_2^{exp\_\theta} - \sigma_2^{theoret\_\theta}. \tag{6}$$

It should be noted that e_1 and e_2 are caused by measuring errors in both coordinate systems (initial and rotated by angle θ). Based on formulas (5)-(6) the error of residual stresses near outer surface (biggest values of stresses) are estimated ~1MPa (near 1%).

2.2 Properties of glass layer – non-destructive testing

In order to perform numerical analysis of glass panel the mechanical properties of the panel should be determined. However, performing traditional mechanical tests with glass specimen is rather complicated, especially tension tests, etc. Furthermore, as mentioned above the non-destructive testing should be preferred. The non-destructive mechanical testing procedure for glass panel was introduced by authors in [18]. The time of flight of the individual pulses has been measured in order to determine sound speeds of P-wave and shear wave. The setup of the test is shown in Figure 2.

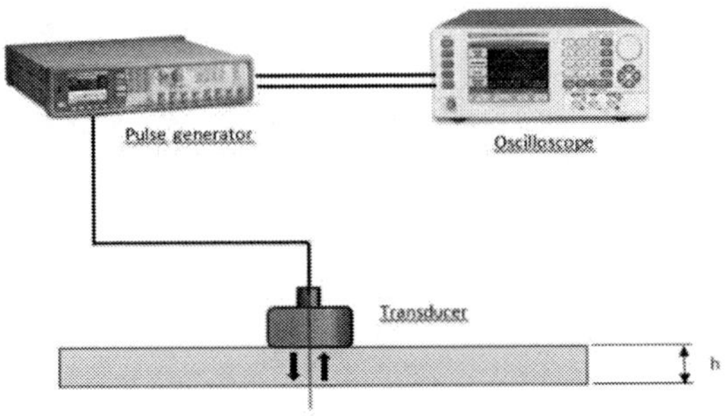

Figure 2. Set up for measuring P-wave and shear wave speeds

The Young modulus and Poisson ratio of the glass can be evaluated as

$$E = \frac{G(3M - 4G)}{M - G}, \tag{7}$$

$$\nu = \frac{M - 2G}{2M - 2G}, \tag{8}$$

where

$$M = c_P^{\,2}\rho, \quad G = c_s^{\,2}\rho, \tag{9}$$

c_P and c_s stand for the wave speeds and ρ is a material density. The wave speeds measured as well as the material properties computed are presented in table 1.

Table 1. Material properties of the glass layer.

	ρ (kg/m^3)	c_P (m/s)	c_s (m/s)	E (GPa)	ν
Float glass	2486	5914	3506	75.11	0.23
Tempered glass	2486	5852	3433	72.53	0.24

The mechanical properties of the PVB interlayer are determined from traditional tensile tests. Due to small thickness and viscoelastic material properties of PVB interlayer the above proposed non-destructive testing is complicated.

3 Numerical modelling

The experimental data acquired by employing procedures introduced in sections 2.1 and 2.2 can be utilized as input data for finite element method (FEM) analysis model. The FE analysis was performed using ANSYS v14.

3.1 FEM model and its validation tests

Modal analysis of glass panels without supports and pre-stress was selected for FE model development and validation. FEA model consists of 50 000 8-noded hexahedral 3D solid elements (8 layers of elements through the thickness of the glass sheet). The FEA and experimental modal analysis has been applied to a glass panel with dimensions 500mm × 200mm × 8mm. The samples are held by a suspension system, constituted by soft elastic bands. These bands are designed to guarantee that the highest rigid mode frequency is less than 10 – 15% of the first resonance frequency of the suspended structure. This circumstance allows considering the systems as in free-free conditions. The classical roving hammer impact test has been used to measure the frequency response functions and to implement the peak-picking method. Preliminarily, a set of points are defined on a wire-frame created on the structure and an accelerometer is fixed on one of these points. The method consists of measuring the frequency response functions between the response measured by the accelerometer and the forced excitation provided by the hammer when the wire-frame points are sequentially hit. In the present investigation, the samples have been schematized with a wire-frame consisting of 12 points displaced in adjacent rectangular patterns. An impulsive excitation has been provided, along the direction perpendicular to the plane of the structure, by means of a hammer AP TechTM AU01 provided with force transducer. The vibration response has been measured, along the same direction, by means of an ICP (Integrated Circuit Piezoelectric) mono- axial accelerometer PCB TM 353B33. The signal acquisition has been performed by a dynamic signal analyser (National Instruments TM NIcDAQ 9174 and NI 9234), controlled by PC based virtual instrument (LabVIEW TM). Random errors have been reduced by averaging out three subsequent measurements taken for each point. The frequency response functions have been estimated as H = SXY /SXX, i.e. as the ratio of the cross-spectrum of excitation and response signals over the auto-spectrum of the excitation signal (Figure 3).

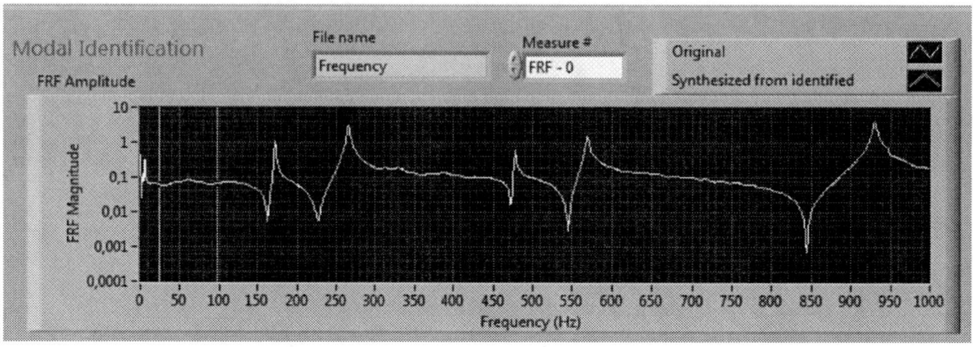

Figure 3. FRF Magnitude vs frequency.

The values of the coherence function related to the impact tests, are above 98%, indicating low levels of uncorrelated noise in the measured frequency response data. In Table 2 the values of the natural frequencies obtained from FEM analysis and experimental tests are presented.

Table 2. Frequency values. FEM vs experimental.

Mode	Frequency, Hz FEM model	Frequency, Hz Experimental
1	172	172
2	267	266
3	477	477
4	571	569
5	930	930

One can be observed from Table 2 that the results obtained from experimental study and FEM analysis are in good agreement.

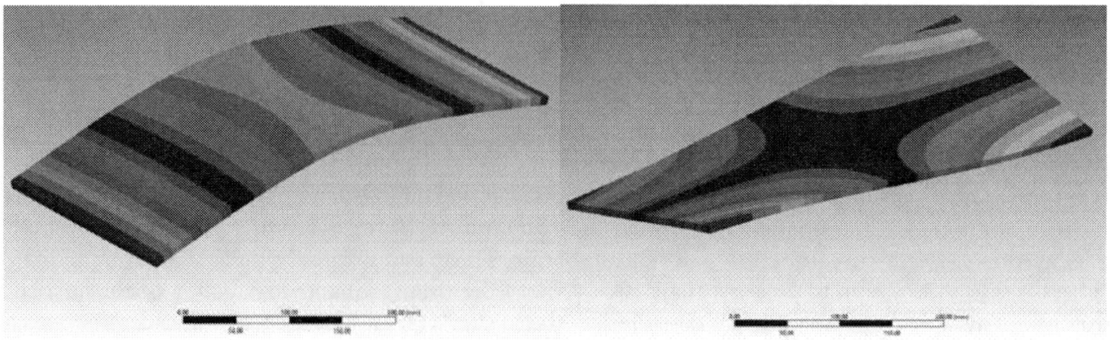

Figure 4. Fundamental frequency and second harmonic.

The fundamental frequency and second harmonic are depicted in Figure 4.

3.2 Modelling residual stresses in glass

Glass panel with size 200x500x8 mm and material properties described above was considered. In order to apply pre-stress one side of panel was fixed (short side is fixed) i.e. console. The values of the stress tensor obtained from experimental study (normal stresses σ_1 and σ_2) and from procedure proposed in section 2.1.2 (shear stress σ_{12}) are mapped into FEA model by ANSYS throughout the thickness in 57834 points (Figure 5).

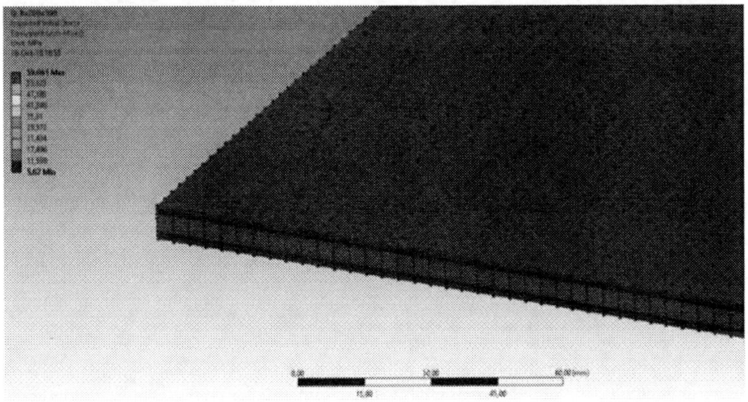

Figure 5. Mapped stresses.

The results of modal analysis are given in Table 3.

Table 3. Frequency values, with and without pre-stress

Mode	Frequency, Hz without pre-stress	Frequency, Hz with pre-stress
1	27	28
2	145	150
3	171	174
4	461	476
5	480	489

In can be observed from Table 3 that in the case of considered boundary conditions (console type, short side fixed) the influence of residual stresses on frequency values is remarkable. Application of residual stresses increase the values of the frequencies. The modes corresponding to first two frequencies are depicted in Figure 6.

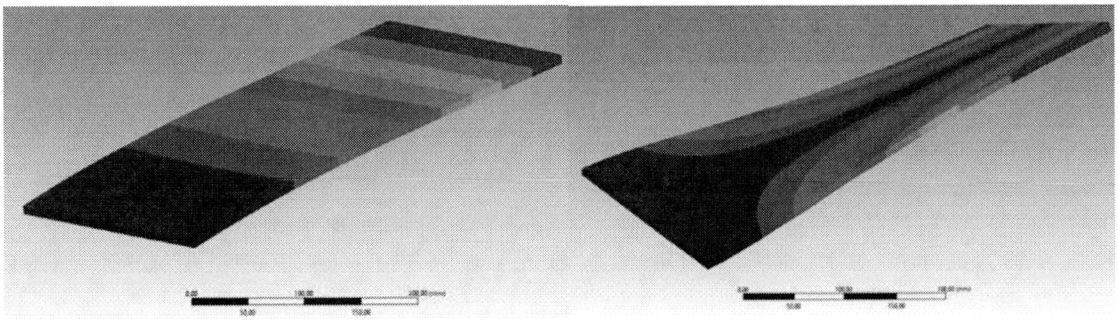

Figure 6. Fundamental frequency and second harmonic for pre-stressed panel.

Figure 6 is obtained by applying console type boundary conditions where short side of the panel is fixed.

4 Conclusions

An approach for structural analysis the glass panels, incorporating residual stresses, has been proposed. The normal residual stresses were determined experimentally in private company GlasStress Ltd. The procedure for determining shear stress component of residual stress tensor has been developed. The algorithm for estimating error of the measured residual stresses has been introduced. The finite element model incorporating residual stresses has been developed. The remarkable influence of the residual stresses on values of the natural frequencies has been observed. An approach proposed for structural analysis can be applied for wide class of glass structures. Future study planned cover design optimization of laminated glass composite panel [19-23].

Acknowledgments

The authors acknowledge Innovative Manufacturing Engineering Systems Competence Centre IMECC (co-financed by the project EU48685); the Estonian Centre of Excellence in Zero Energy and Resource Efficient Smart Buildings and Districts, ZEBE, TK146 funded by the European Regional Development Fund (grant 2014-2020.4.01.15-0016); Estonian Research Council grant PUT1300.

References

1. M. Haldimann, A. Luible A and M. Overend, Structural Engineering Documents, 10. IABSE, Zürich (2008).
2. M. Kotoul, J. Pokluda, P. Sandera, I. Dlouh?, Z. Chlup, A.R. Boccaccini, Acta Mater., **56**, 2908-2918 (2008).
3. F.C. Serbena, E.D. Zanotto, Journal of Non-Crystalline Solids, **358**, 6-7, 975-984 (2012).
4. M. Fröling, *Strength Design Methods for Glass Structures*, Doctoral Thesis (2013).
5. X. Zhang H. Hao, Construction and Building Materials, **93**, 404-415 (2015).
6. F. Auriemma, H. Rammal, J. Lavrentiev, SAE Int. J. of Materials and Manufacturing, **1**,1, 600-611 (2014).
7. F. Auriemma, Applied Acoustics,**122**, 128-137 (2017).
8. F. Auriemma, Acoustics Australia, **45**, 2, 411-419 (2017).
9. F. Auriemma, Proc. of Meetings on Acoustics Meeting of Acoust. Society of America, **30**, 1 (2017).
10. S. Hödemann, A. Valdmann, J. Anton, T. Murata, J. of Materials Sci., **51**,12 (2016).
11. H. Aben, J. Anton, M. Õis, K. Viswanathan, C. Chandrasekar, M. M. Chaudhri, App. Physics Letters, **109**, 231903 (2016).
12. H. Aben, D. Lochegnies, Y. Chen, J. Anton, M. Paemurru, M. Õis, Experim. Mech., **55**, 2, 483-486 (2015).
13. J. Lellep, J. Majak, Struct. Optimization, **14**, 116-120 (1997).
14. J. Lellep, J. Majak, Mech. of Composite Materials, **36**, 4, 261-266 (2000).
15. A. Aruniit, J. Kers, J. Majak, A. Krumme, K. Tall, Proc. of Estonian Acad. of Sci., **61**, 3, 160-165 (2012).
16. A. Aruniit, J. Kers, D. Goljandin, M. Saarna, K. Tall, J. Majak, H. Herranen, Materials Science (Medžiagotyra), **17**, 3, 276-281 (2011).
17. J.Majak, S.Hannus, Mech. of Composite Materials, **39**, 6, 509-520 (2003).
18. K.Väer, J. Anton, A. Klauson, M. Eerme, E. Õunapuu, P. Tšukrejev, Journal of Achievements in Materials and Manufacturing Engineering, **81**,1, 11−17 (2017).
19. J. Šliseris, K. Rocens, World Acad. of Sci., Eng. and Tech., 76, 317-323 (2011).
20. J. Šliseris, K. Rocens, J. of Civil Eng. and Management, 19 (5), 696-704 (2013).
21. W.H.Zhang, M. Domaszewski, D. Bassir, Structural Optimization, **17**, 2, 219-225 (1999).
22. S. Guessasma, D. Bassir, L. Hedjazi, Materials and Design, **65**, 1053-1063 (2015).
23. J.Zhu, W. Zhang, L. Xia, Q. Zhang, D. Bassir, Journal of Intelligent and Robotic Systems: Theory and Applications. **67**, 3-4, 185-199 (2012).

MATEC Web of Conferences **253**, 02004 (2019) https://doi.org/10.1051/matecconf/201925302004
MSME 2018

Effect of Fiber Volume Fraction on Behavior of Concrete Beams Made with Recycled Concrete Aggregates

Wael Alnahhal[1,a] and Omar Aljidda[2]

[1]*Department of Civil and Architectural Engineering, College of Engineering, Qatar University, Qatar*
[2]*Department of Civil and Water Engineering, Laval University, Canada*

Abstract. This study investigates the effect of using different volume fractions of basalt macro fibers (BMF) on the flexural behavior of concrete beams made with 100% recycled concrete aggregates (RCA) experimentally. A total of 4 reinforced concrete (RC) beam specimens were flexural tested until failure. The parameter investigated included the BMF volume fraction (0%, 0.5%, 1%, and 1.5%). The testing results of the specimens were compared to control beam specimen made with no added fibers. The experimental results showed that adding BMF improves the flexural capacity of the tested beams.

1 Introduction

Previous studies have shown that using discrete steel fibers in concrete increases its ductility due to the large compressive strains exhibited at failure and the long-term severability [1-3]. However, among the disadvantages of using steel fibers is corrosion, especially in a harsh environment that characterizes the Arabian Gulf region, and the added weight to the concrete structure. Therefore, basalt macro-fibers were proposed in this study. Basalt fiber is a natural material that is found in volcanic basalt rocks formed in a melting temperature comprised between 1500 to 1700 C. Basalt fibers are considered to be one of the relatively new composite materials, which are characterized by their high corrosion and thermal resistance [4-6]. Basalt macro-fibers possess higher tensile strength and higher stiffness compared to conventional synthetic fibers [7]. Basalt macro-fibers modify the cracking mechanism in concrete which results in an increase in the flexural tensile strength and average residual strength, and a decrease in the crack width of the concrete [7-9]. Krassowska [10] and Brik [11] investigated the mechanical characteristics of fibers depend on their length and diameter. Based on their study, fibers with smaller diameter showed a higher tensile strength and modulus of elasticity than those fibers with a larger diameter. Jiang et al. [12] reported an improvement in the engineering properties of fiber reinforced concrete when both the volume fraction and length of fibers were increased. Bajaj [13] reported that the crack width that develops in plain concrete is larger than the crack width in polypropylene fiber concrete, which is also larger than the crack width obtained in reinforced concrete having chopped basalt fibers in the mix. Tumadhir [14] indicated that increasing the volume fraction of basalt fibers resulted in increasing the splitting tensile strength of concrete with no remarkable effect on the compressive strength of concrete. Das [15] indicated that the optimum percentage of basalt fibers in a concrete mixture is from 0.5% to 0.8% of cement's weight in order to

[a] Corresponding author : wael.alnahhal@qu.edu.qa

achieve the best concrete properties. Dias and Thaumaturge [16] and Dong et al. [17] focused on the mechanical properties of the concrete made with RCA and basalt fibers. The results obtained from this study showed that the mechanical properties of concrete made with RCA are decreased with increasing the RCA replacement ratio. However, the mechanical properties were enhanced by using the basalt fibers. A few studies examined the flexural behavior of FRC beams. Byung and Oh [18] investigated the flexural behavior of steel FRC beams. The results indicated that the ultimate strength, ductility, and energy absorption capacity of FRC beams were considerably enhanced with the addition of steel fibers. Furthermore, the crack width was remarkably reduced as the amount of fibers increased. Similar test results were reported by Mertol et al.[19]. This work attempts to investigate the feasibility of using different volume fractions of BMF combined with 100% of RCA for structural application. Thus, this study discusses the experimental results for the flexural testing of large-scale basalt FRC beams made with 100%.

2 Experimental program

The experimental program aimed at investigating the effect of using different volume fractions of BMF on the flexural behavior of concrete beams made with 100% recycled concrete aggregates (RCA) under four-point loading.

2.1 Concrete mixture proportions

The coarse aggregates used in this study were RCA with a maximum size of 20 mm. The water/cement ratio of the mix was 0.45. The target compressive strength of concrete was 35 MPa. As the RCA is characterized by its higher water absorption due to the attached old mortar, the amount of free water in concrete mixture plays a significant rule in the development of the hardened properties for resulting concrete. For this reason, RCA was washed and immersed in water for 24 hours before mixing, and then the surface moisture was dried with a moistened cloth, as shown in Figure1.This is to ensure that both aggregates were in a saturated surface dry condition at the time of concrete mixing. Water absorbed by saturated aggregates was not included in calculating water to cement content. This is to ensure that the remaining amount of water is approximately the same and is enough to ensure the hydration of the cement particles in the concrete mix.

BMF with an average diameter 0.72 mm and a length of 45 mm was used in this study. It is a non-corrosive discrete thin fiber made from basalt stone and coated with a solution suitable for use in concrete. It possesses higher tensile strength and stiffness, It acts as a proactive reinforcement that provides the immediate tensile load carrying capacity when micro-cracks develop in concrete. It has a tensile strength of 900 MPa and a modulus of elasticity of 44 GPa.

2.2 Large-scale beams flexural tests

To study the flexural behavior of beams made with different volume fractions of BMF and with100% of RCA, the test was carried on four large-scale beams with a size of 150× 250 × 2550 mm that were loaded until failure under four-point loads as shown in Figure 2. Steel bars were used for as the main flexural steel reinforcement. 8-mm-diameter steel stirrups were used as transverse steel reinforcement, and two 16-mm-diameter steel bars were used as the main longitudinal bottom reinforcement. Steel stirrups were spaced as shown in Figure 3 in order to have a flexural failure at midspan of each beam. The concrete clear cover was 25 mm.

Figure 1. Moisturizing and Drying of RCA.

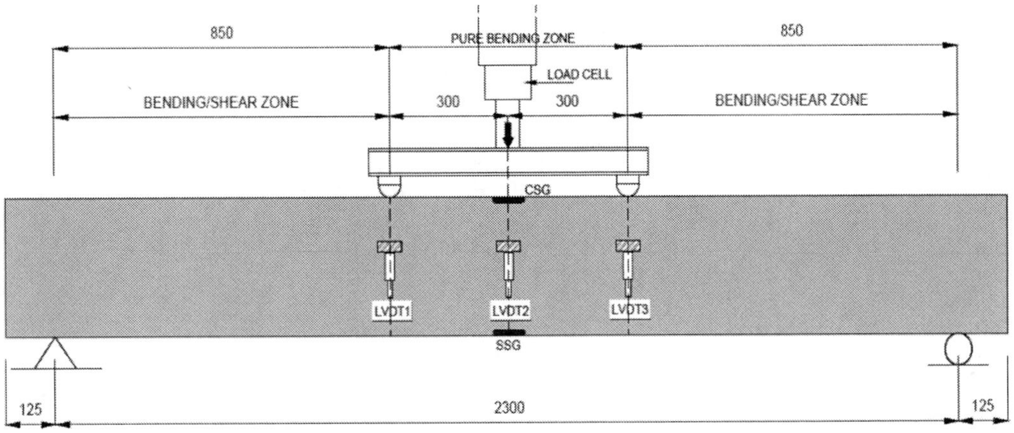

Figure 2. Large-Scale Beam Layout (Dimensions in mm).

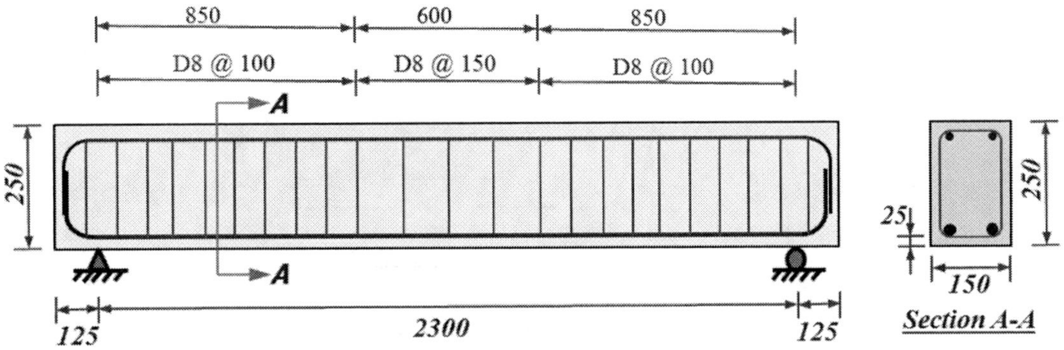

Figure 3. Beam Details and Cross Section (Dimensions in mm).

3 Figures and tablesTests results and discussion

All four beams specimens have 100% of RCA with 0%, 0.5%, 1%, and 1.5% of BMF respectively. Flexure failure at the mid-span was the main failure mode for all four tested beams. As expected, the addition of BMF to concrete increased the maximum deflection at failure about 11.76%, 34.78% and 44.44% for beams with the volume fraction of 0.5%, 1%, and 1.5% respectively compared with control beam as shown in Figure 4(a). The reason can be due to the effects of BMF and high bonding between BMF and concrete matrix. Figure 4(b) shows the load-concrete strain relationship where the maximum concrete strain values were 0.00292, 0.00308, 0.0035 and 0.00372 for beams with 0%,0.5%,1% and 1.5% BMF respectively. The differences of the load-strain curves between the control beam and beams with BMF are significant. In control beam that has 0% BMF, once reaching the ultimate load, concrete failed by crushing and strain in reinforcement dropped suddenly. however, when the beams with BMF reached the ultimate load, concrete was held together and the strains in concrete and steel bars kept increasing gradually, this is due to the behavior of BMF, tensile loads can be transferred across the cracks by the bridging of fiber and increase in strain capacity of the concrete matrix in the pre-failure zone.

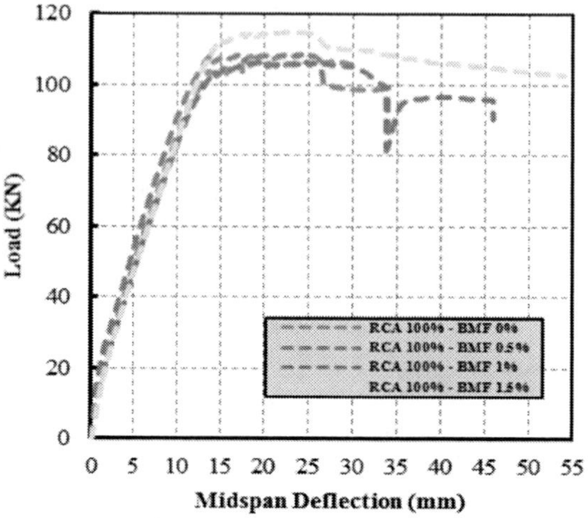

a.) Load- Midspan Deflection

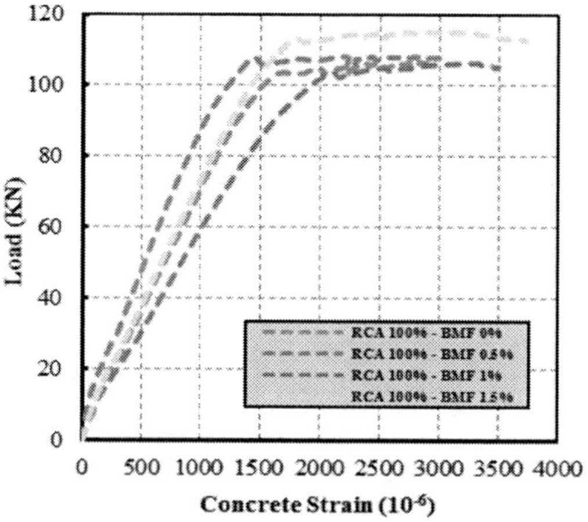

b.) Load- Concrete Strain

Figure 4. Comparison of Test Results for Beams with 100% of RCA and different volume fraction of BMF

4 Conclusions

This study has investigated the mechanical properties and flexural behavior of RC beams with the different volume fraction of BMF and 100% of RCA. The following conclusions can be drawn based on the findings of the experimental investigation:

1. The addition of BMF resulted in an improvement in the ultimate flexural capacity of the tested beams compared to the beams with no BMF.

The addition of BMF resulted in an improvement in the ductility of the tested beams compared to the beams with no BMF.

Acknowledgment

The authors show their gratitude to Qatar University for their financial support through the internal research grant QUUS-CENG-SPR14/15-21. The authors are grateful to the Construction Materials Co L.L.C. (Qatar Quarry Company) for providing the RCA.

References

1. ACI Comite 544, State of the Art Report on Fiber Reinforced Concrete Reported (ACI 544.1R-96 Reapproved 2002). ACI Structural Journal, 96(Reapproved), p.66 (2002)
2. K. Holschemacher, T. Mueller, and Y Ribakov, Effect of steel fibers on mechanical properties of high-strength concrete. Materials and Design, **31**(5), pp.2604–2615 (2010)
3. J. Katzer and J. Domski, Quality and mechanical properties of engineered steel fibers used as reinforcement for concrete. Construction and Building Materials, **34**, pp.243– 248 (2012)
4. A. El Refai, F. Abed, and M. Asce, Concrete Contribution to Shear Strength of Beams Reinforced with Basalt Fiber-Reinforced Bars, Journal of Composites for Construction, vol. **20**, no. 4 (2016)
5. W. Alnahhal, and O. Aljidda, Flexural behavior of basalt fiber reinforced concrete beams with recycled concrete coarse aggregates. Construction and Building Materials, **169**, 165-178 (2018)
6. J. Sim, C. Park, and D. Moon, Characteristics of basalt fiber as a strengthening material for concrete structures, Composites Part B: Engineering, vol. **36**, pp. 504–512 (2005)
7. S. Adhikari, Mechanical and Structural Characterization of Mini-Bar Reinforced Concrete Beams. Thesis (2013)
8. P. Anil, M.Len, and A. Sudeep, Basalt Frp Minibar Reinforced Concrete, Fibre Concrete, pp. 12–13 (2013)
9. P. Banibayat, and Patnaik: Variability of Mechanical Properties of Basalt Fiber Reinforced Polymer Bars Manufactured By Wet-Layup Method. Materials and Design, **56**, pp.898–906. (2014)
10. J. Krassowska and A. Lapko, The Influence of Steel and Basalt Fibers on the Shear and Flexural Capacity of Reinforced Concrete Beams. Journal of Civil Engineering and Architecture. vol. **7**, no. 7, pp. 789–795 (2013)
11. V. Brik, Basalt Fiber Composite Reinforcement for Concrete, Transportation Research Board National Research Council, Madison, Wisconsin (1997)
12. C. Jiang, K. Fan, F. Wu, and D. Chen, Experimental study on the mechanical properties and microstructure of chopped basalt fibre reinforced concrete: Mater. Des., vol. **58**, pp. 187–193, (2014)
13. S. Bajaj, Effect of Corrosion on Physical and Mechanical Properties of Reinforced Concrete. Ph.D. Thesis (2012)
14. M. Tumadhir, Thermal and Mechanical Properties of Basalt Fiber Reinforced Concrete, World Academy of Science, Engineering and Technology, vol. **7**, no. 4, pp. 712–715 (2013)
15. W. Das, The impact of basaltic fibre on selected physical and mechanical properties of cement mortar. Concert Precast. Plant Technology, vol. **71**, no. 3, pp. 28–33 (2005)
16. D. Dias and C. Thaumaturgo, Fracture toughness of geopolymeric concretes reinforced with basalt fibers: Cement and Concrete Compost, vol. **27**, no. 1, pp. 49–54 (2005)
17. J. Dong, Q. Wang, and Z. Guan, Material properties of basalt fibre reinforced concrete made with recycled earthquake waste, Construction and Building Material, vol. **130**, pp. 241–251 (2017)
18. B. Byung and H. Oh: Flexural Analysis of Reinforced Concrete Beams Containing Steel Fibers, Journal of structural engineering, vol. **118**, issue10, pp. 2821–2835 (1992)
19. H. Mertol, E. Baran, and H. Bello, Flexural behavior of lightly and heavily reinforced steel fiber concrete beams, Construction and Building Material, vol. **98**, pp. 185–193 (2015)

Recycling of Mixed Poly(Ethylene-terephthalate) and Poly(Lactic Acid)

Daniel GERE[1] and Tibor CZIGANY[1, 2, a]

[1]*Department of Polymer Engineering, Faculty of Mechanical Engineering, Budapest University of Technology and Economics, H-1111 Budapest, Muegyetem rkp. 3, Hungary*
[2]*MTA-BME Research Group for Composite Science and Technology, H-1111 Budapest, Muegyetem rkp. 3., Hungary*

Abstract. Nowadays, PLA is increasingly used as a packaging material, therefore it may appear in the petrol-based polymer waste stream. However, with the today's mechanical recycling technologies PLA and PET bottles cannot be easily or cheaply separated. Therefore, our goal was to investigate the mechanical, morphological and thermal properties of different PET and PLA compounds in a wide range of compositions. We made different compounds from poly(ethylene-terephthalate) (PET) and poly(lactic acid) (PLA) by extrusion, and injection molded specimens from the compounds. We investigated the mechanical properties and the phase morphology of the samples and the thermal stability of the regranulates. PET and PLA are thermodynamically immiscible, therefore we observed a typical island-sea type morphology in SEM micrographs. When PLA was added, the mechanical properties (tensile strength, modulus, elongation at break and impact strength) changed significantly. The Young's modulus increased, while elongation at break and impact strength decreased with the increase of the weight fraction of PLA. The TGA results indicated that the incorporation of PLA decreased the thermal stability of the PET/PLA blends.

1 Introduction

More and more polymers are produced and processed annually but biopolymers still only amount to slightly less than 1.5 wt% of all polymers produced [1, 2]. Biron [3] predicted that by 2040 the same amount of biopolymers will be produced as petroleum-based polymers and then more biopolymers will be produced than petroleum-based polymers. Most petroleum-based polymers and biopolymers are used to produce packaging materials, which become waste in a very short time [1, 2, 4]. There are several ways to recycle polymers—mechanical and chemical recycling, and in the case of biopolymers, also biological recycling [5]. The population and selective waste collection are not prepared to sort these polymers, therefore they can mix during recycling. The question is what happens when they are processed together. How do their mechanical, morphological and rheological properties change?

There are publications about blends of petroleum-based and biopolymers but some of these [6-8] aimed to improve the unfavourable properties of PLA (e.g. rigidity, low tolerance of heat) [9, 10], usually with the addition of polyethylene. The blends created this way are immiscible, therefore other

[a] Corresponding author: czigany@eik.bme.hu

additives are necessary. Only a few authors examined the effect PLA as a „contaminant" [11-13], and only in a very narrow range.

For example, PLA and PET bottles cannot be easily or cheaply sorted, because both are transparent and very similar, therefore manual sorting based on visual appearance is not possible. The density of PET and PLA are similar, higher than water, therefore in the float-sink separation process, the PLA particles will follow the PET flakes to the bottom of the flotation tank. Therefore, the PLA fragments and the PET waste stream will be recycled together [14, 15]. The most widely used technique for the automated sorting of plastics [16] is Fourier Transform near-infrared (FT-NIR) spectroscopy. However the reports [14] estimate that the efficiency of the separation of PLA bottles from PET bottles is 86–99.6 % for this method.

The presence of biopolymers can significantly modify the properties of PET [11, 17]. The processing temperature of PET is around 100 °C higher than that of PLA therefore PLA starts to degrade, which leads to a yellowing of the product. Additionally, PLA has a different glass transition temperature, which results in opaqueness or haziness when PET and PLA are processed together. PLA and PET are thermodynamically immiscible, therefore holes, peaks or clusters will probably occur in the products [14].

The analysis of statistical data [1, 2] indicates that nowadays it is reasonable to examine the effect of low biopolymer "contamination" (<5 wt%). We, however, are of the opinion that as biopolymers are more and more widely used, the 20 wt%–80 wt% range should also be examined for the future, as should the effect of a small amount of petroleum-based polymer contamination of biopolymers.

Therefore, our goal was to investigate the mechanical, morphological and thermal properties of different PET and PLA compounds in a wide composition range.

2 Experimental

2.1 Materials

We used virgin PET type NeoPET 80 (intrinsic viscosity (IV): 0.80 dl/g, density: 1.34 g/cm^3) supplied by NeoGroup (Lithuania) as a petroleum-based component, and virgin PLA type Ingeo 3100HP (MFR (210°C, 2.16 kg): 24 g/10 min, density: 1.24 g/cm^3), supplied by NatureWorks LLC. (USA), as a biobased component.

2.2 Material preparation and processing

Before processing (extrusion and injection molding), the materials were dried at 140°C in a Faithful WGLL-125 BE hot air drying oven for 6 hours. Thirteen different PET/PLA compounds (regranulates) (100/0, 99/1, 97/3, 95/5 85/15, 75/25, 50/50, 25/75, 15/85, 5/95, 3/97, 1/99, 0/100 wt/wt%) were produced with a Labtech Scientific LTE 26-44 co-rotating twin-screw extruder. The screw diameter of the extruder was 26 mm and the length/diameter (L/D) ratio was 44. The temperature profile of the extruder (from intake zone to die) was 235-240-245-250-255-260-265-270-275-270-265°C. The rotating speed of the extruder screws was 50 rpm and the melt pressure was 30-50 bar.

The injection molded samples were made with an Arburg Allrounder 370 S 700-290 injection molding machine. The injection rate was 50 cm^3/s, holding pressure was 800 bar, holding time was 20 s, residual cooling time was 40 s, and melt and mold temperature were 280°C and 30°C, respectively. The mechanical tests were carried out on ISO standard dumbbell-shaped tensile specimens.

2.3 Methods

Tensile tests were carried out on a Zwick Z005 universal testing machine. We calculated the tensile modulus between 0.05% and 0.25% strain with a crosshead speed of 1 mm/min, and determined tensile strength, breaking strength, elongation at the maximum force and elongation at break with a crosshead speed of 10 mm/min. The measurements were performed on ISO 527-2/1A dumbbell-shaped specimens with an overall length of 170 mm and a cross-section of 4 mm × 10 mm. In every measurement, we tested 5 specimens, and calculated the average value and standard deviation.

Impact strength was determined with the Charpy impact test on a Ceast Resil Impactor Junior impact tester, with a pendulum of 2 J. The measurements were performed on 2 mm notched ISO 179-1/1eA specimens with a length of 80 mm and a cross-section of 4 mm × 10 mm. The tests were performed at room temperature and at a relative humidity of 50%. In every measurement, we tested 10 specimens, and calculated the average and standard deviation.

The morphology of the blends was studied with a scanning electron microscope (Jeol JSM 6380LA). The SEM micrographs were taken from cryo-fractured surfaces.

Thermogravimetric analysis (TGA) was performed on samples of about 10 mg with a TA Instruments Q500 instrument over 40-600°C in an industrial grade air (78% N_2, 21% O_2, 1% other) atmosphere (60 ml/min) with a 10 °C/min heating speed.

3 Results and discussion

3.1 Mechanical properties of the PET/PLA blends

Figure 1 shows the tensile stress-strain curves of the PET and the PLA. The curves show that the PLA was rigid and PET was flexible after the neck formation.

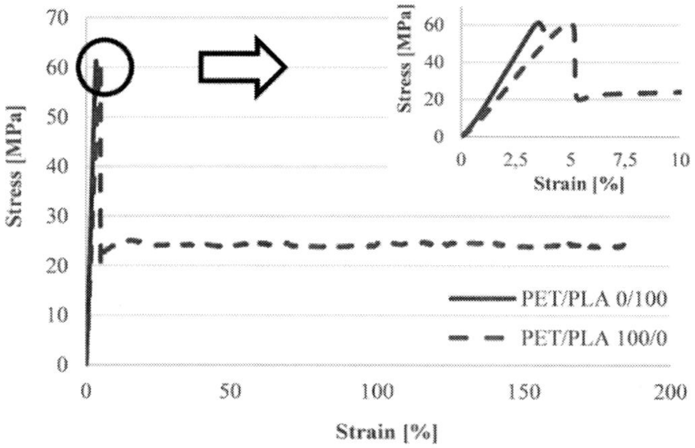

Figure 1. Tensile stress-strain curves of 100/0 and 0/100 PET/PLA samples

Figure 2 shows the mechanical properties of the injection molded samples of the different PET/PLA blends. The tensile strength values were similar in the case of small (≤25 wt%) and large (≥85 wt%) PLA content. However, it showed a lower value in 50/50 PET/PLA composition. When the weight fraction of PLA was increased, the elongation at the maximum force did not change initially, but then gradually decreased and was finally close to a constant value. Breaking strength in the case of low PLA content (≤25 wt%) was lower than in the other cases. It can be explained by the fact that the specimens were broken after the neck formation when the PLA content was small, while in all other cases, they were broken rigidly. In the case of low PLA content (≤5 wt%), elongation at break did not change, compared to pure PET, but when the weight fraction of PLA was further increased, it decreased sharply and stayed low. In addition to the presence of a rigid component (PLA), the degradation of PLA or both PLA and PET, and the increased droplet size (see Figure 3a–d) may also have led to a decrease in elongation at break when the weight fraction of PLA was increased. We observed that the Young's modulus increased linearly, according to the linear rule of mixtures, with the increase of the weight fraction of PLA. On the other hand, as the weight fraction of PLA was increased, Charpy impact strength seemed to change along a parabola. 1 wt% PET or PLA contamination did not affect Charpy impact strength, but in all other cases, it decrease significantly compared to the pure raw materials. This is due to the poor adhesion between the PET and the PLA, and to the fact that dispersed droplets are present in the blends as failures. La Mantia *et al.* [11] had similar results when they investigated the 99.5/0.5, 99/1, 98/2 and 95/5 PET/PLA blends.

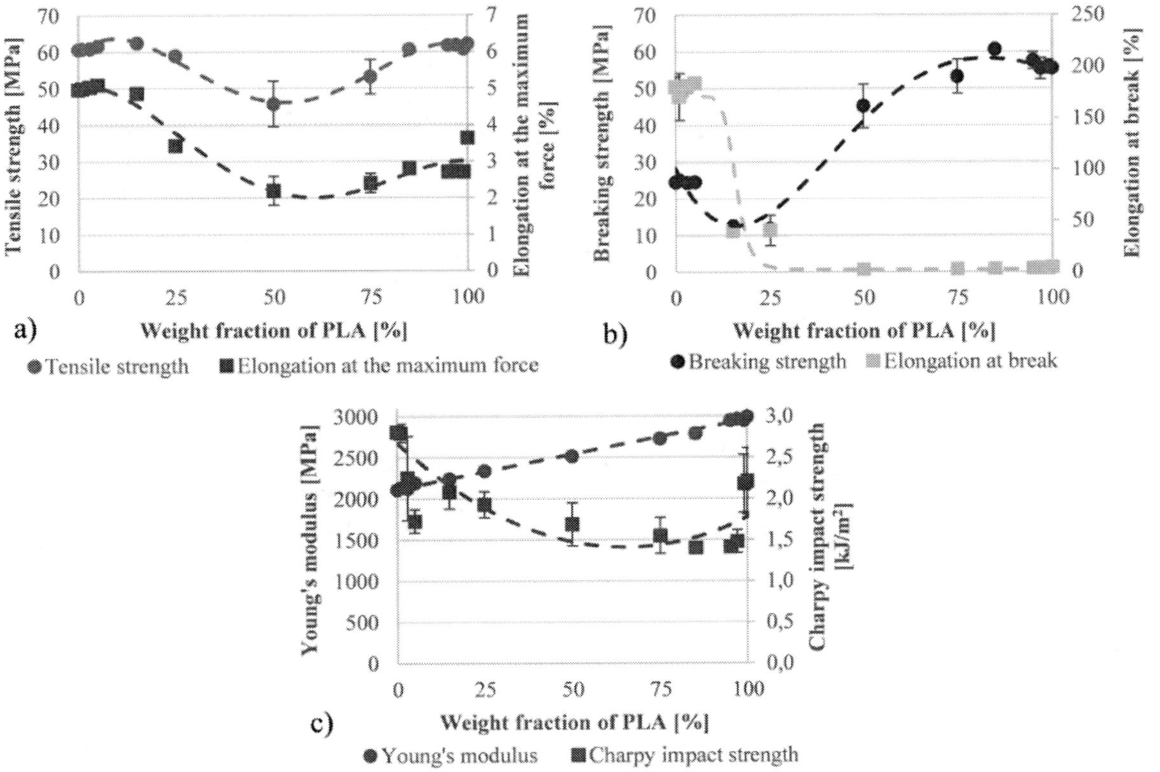

Figure 2. Tensile strength and elongation at the maximum force (a), breaking strength and elongation at break (b), Young's modulus and Charpy impact strength (c) in PET/PLA blends depending on the weight fraction of PLA

3.2 Morphological structure of the blends/Phase morphology of the blends

We also studied the structure of the blends by SEM (Figure 3). The micrographs were taken from cryo-fractured surfaces. The SEM micrographs indicated that a dispersed phase structure (island-sea type morphology) was formed for each composition, only the droplet size and the matrix were different. The SEM micrographs showed a typical morphology of immiscible polymer blends if one of the components had a well-defined spherical shape, very broad size distribution and a smooth fracture surface. As PLA content was increased, up to 25 wt% of PLA, the diameter of the dispersed PLA particles in the PET matrix increased from a few tenths of a micron to a few microns. In the case of a symmetrical composition (50 wt% of PET and PLA), it is not possible to determine which phase constitutes the matrix and which is the dispersed phase. However, according to the literature [18], in this case the polymer with lower viscosity forms the matrix, in our case the PLA. When the weight fraction of the PLA was further increased, it formed the matrix and PET was the dispersed phase. Torres-Huerta *et al.* [12] also observed a typical island-sea type morphology in PET/PLA (95/5, 90/10 and 85/15) blends.

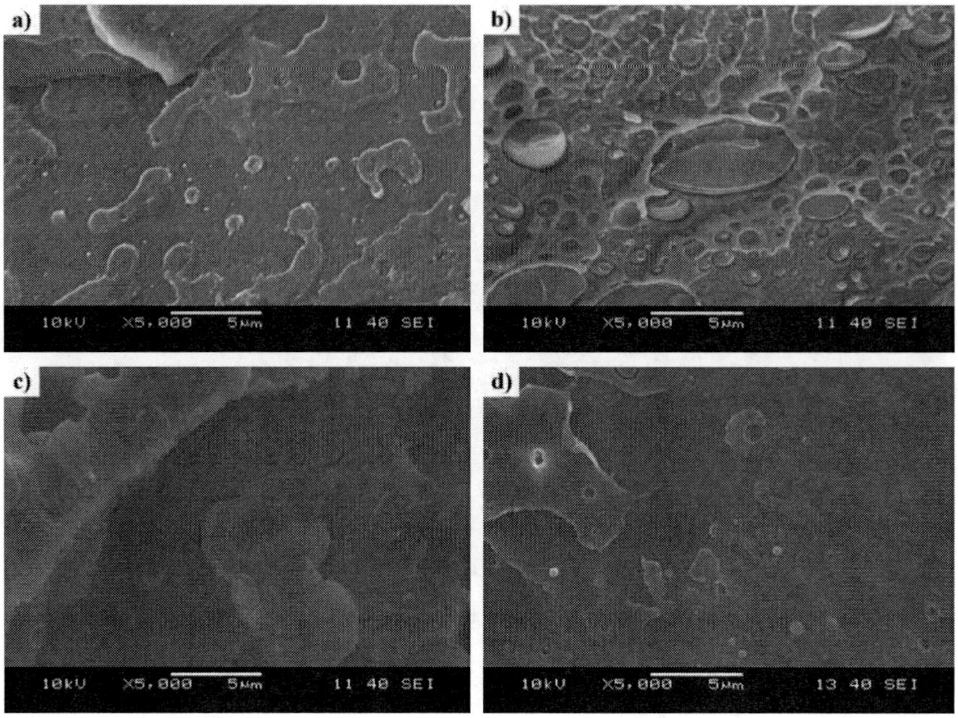

Figure 3. Changes in the morphology of PET/PLA blends a) 1 wt% PLA, b) 25 wt% PLA, c) 50 wt% PLA, d) 95 wt% PLA

3.3 Thermal stability of the blends

The TGA and the corresponding first derivative TGA (DTG) curves of PET/PLA blends are shown in Figure 4. The degradation onset temperature (the temperature at which 5 wt% degradation occurred) of the 100/0, 99/1, 97/3, 95/5 85/15, 75/25, 50/50, 25/75, 15/85, 5/95, 3/97, 1/99 and 0/100 PET/PLA blends were 371 °C, 369 °C, 364 °C, 358 °C, 347 °C, 342 °C, 324 °C, 324 °C, 315 °C, 318 °C, 310 °C, 315 °C and 318 °C, respectively. The maximum degradation temperature (the peak temperature of the DTG curve) of the 100/0, 99/1, 97/3, 95/5 85/15, 75/25, 50/50, 25/75, 15/85, 5/95, 3/97, 1/99 and 0/100 PET/PLA blends were 426 °C, 416 °C, 424 °C, 425 °C, 424 °C, 424 °C, 354 °C, 358 °C, 355 °C, 359 °C, 353 °C, 351 °C and 360 °C, respectively. The results indicated that the thermal degradation of the PET/PLA blends has two maximum degradation temperatures, about 355 °C and 424 °C, corresponding to PLA and PET, respectively. These results show that adding PLA to PET decreased the thermal stability of the blends. McLauchlin and Ghita [19] had similar results in a nitrogen atmosphere when they investigated PET/PLA blends which contained 0.5, 1, 2, 5, 10 and 20 wt% of PLA.

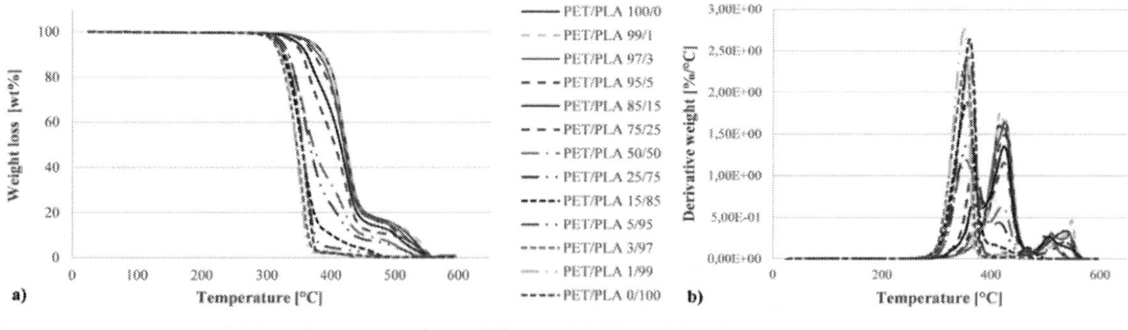

Figure 4. TGA (a) and DTG (b) curves of the different PET/PLA blends

4 Conclusions

We investigated the mechanical and morphological properties, and the thermal stability of different PET/PLA blends. The stress-strain curves showed that PET was elastic and PLA was rigid. As the weight fraction of PLA was increased from 0% to 100%, the Young's modulus increased from 2109 MPa to 2990 MPa, while elongation at break decreased from 180 % to 4 %. The tensile strength values were similar in the case of low (≤25 wt%) and high (≥85 wt%) PLA content. In SEM micrographs, we observed a typical island-sea type morphology in all the PET/PLA blends. Therefore, the two components are thermodynamically incompatible, at least in the concentrations investigated. The thermal stability of the blends was decreased when PLA was added to the blends. The result indicate that blends with more than 3 wt% PLA contamination or more than 1 wt% PET contamination would probably result in a lot of faulty products, due to the immiscibility of the components.

Acknowledgements

This paper was supported by the National Research, Development and Innovation Office (grant number: NVKP_16-1-2016-0012) and by the Higher Education Excellence Program of the Ministry of Human Capacities in the frame of Nanotechnology research area of Budapest University of Technology and Economics (BME FIKP-NANO).

References

1. PlasticsEurope, *Plastics – the Facts 2016* (PlasticsEurope AISBL, Brussels, 2017)
2. European Bioplastics, *Bioplastics packaging – combining performance with sustainability* (European Bioplastics e.V., Berlin, 2017)
3. M. Biron, *Industrial applications of renewable plastics: environmental, technological, and economic advances* (William Andrew, Oxford, 2017)
4. R. Geyer, J. R. Jambeck, K. L. Law, Sci. Adv., **3**, e1700782 (2017)
5. A. Soroudi, I. Jakubowicz, Eur. Polym. J., **49**, 2839-2858 (2013)
6. X. Lu, L. Tang, L. Wang, J. Zhao, D. Li, Z. Wu, P. Xiao, Polym. Test., **54**, 90-97 (2016)
7. N. Yarahmadi, I. Jakubowicz, J. Enebro, J. Appl. Polym. Sci., **133**, 43916 (2016)
8. Y. F. Kim, C. N. Choi, Y. D. Kim, K. Y. Lee, M. S. Lee, Fiber. Polym., **5**, 270-274 (2004)
9. C. Vasile, M. Râpă, M. Ştefan, M. Stan, S. Macavei, R. N. Darie-Nită, L. Barbu-Tudoran, D. C. Vodnar, E.E. Popa, R. Ştefan, G. Borodi, M. Brebu, Express Polym. Lett., **11**, 531-544 (2017)
10. W. S. Chow, Y. Y. Leu, Z. A. M. Ishak, Period. Polytech. Mech. Eng., **60**, 15-20 (2016)
11. F. P. La Mantia, L. Botta, M. Morreale, R. Scaffaro, Polym. Degrad. Stabil., **97**, 21-24 (2012)
12. A. M. Torres-Huerta, D. Palma-Ramírez, M. A. Domínguez-Crespo, D. Del Angel-López, D. de la Fuente, Eur. Polym. J., **61**, 285-299 (2014)
13. D. Gere, T. Czigany, Waste Manage., **76**, 190-198 (2018)
14. L. Alaerts, M. Augustinus, K. Van Acker, Sustainability, **10**, 1487 (2018)
15. M. R. Gent, M. Menendez, J. Torano, I. Diego, Waste Manage. Res., **27**, 175-187 (2009)
16. J. Hopewell, R. Dvorak, E. Kosior, Philos. Trans. R. Soc. B-Biol. Sci., **364**, 2115-2126 (2009)
17. D. D. Cornell, J. Polym. Environ., **15**, 295-299 (2007)
18. J. K. Lee, C. D. Han, Polymer, **41**, 1799-1815 (2000)
19. A. R. McLauchlin, O. R. Ghita, J. Appl. Polym. Sci., **133**, 44147 (2016)

Antibacterial Effect of Ginseng/polyaniline Encapsulated in Poly(Lactic-*co*-glycolic Acid) Microcapsules Coating on Stainless Steel 316L

Siti Khadijah Lukman[1] and Syafiqah Saidin[1,2,a]

[1]*School of Biomedical Engineering & Health Sciences, Faculty of Engineering, Universiti Teknologi Malaysia, 81310 UTM Johor Bahru, Johor, Malaysia.*
[2]*IJN-UTM Cardiovascular Engineering Centre, Institute of Human Centered Engineering, Universiti Teknologi Malaysia, 81310, UTM Johor Bahru, Johor, Malaysia.*

Abstract. Implant infection is one the current complications of implant restoration. Incorporation of a material that possesses antibacterial property is crucial in combating implant infection. In this study, polyaniline (PANI) is incorporated in ginseng encapsulated poly (lactic-*co*-glycolic acid) (PLGA) microcapsules to equip the microcapsules with an antibacterial effect. The microcapsules are intended for drug delivery purpose as the ginseng is comprised of abundant therapeutic values and the PLGA is known for its degradation property. The microcapsules were coated on stainless steel 316L using an electro-deposition technique. The chemical composition, morphology and antibacterial activity of the samples were analyzed through ATR-FTIR, SEM and bacterial count test. A greater volume of microcapsules was coated on the stainless steel 316L with the addition of PANI through the appearance of intense ginseng (O–H and C=C bands) and PLGA (C=O) peaks on the ATR-FTIR spectra and through the visualization of spherical morphology of microcapsules on the pre-treated metal. The incorporation of PANI also has induced the antibacterial efficacy of the microcapsules to 87.64%. Therefore, PANI has served as an antibacterial agent that could be useful in the development of implant coating materials while driving the formation of ginseng encapsulated PLGA microcapsules.

1 Introduction

Implant is a temporary or permanent substitute to replace dysfunction organs. Despite of continuous advancement in metallic implant techniques and materials, implant infection is among the current complications of implant restoration in various medical fields such as dentistry [1], cardiovascular [2] and orthopedics [3]. The treatment of implant infection is time-consuming and costly which might lead to the replacement of failure infected implants [3]. Hence, providing an antibacterial property on implant materials is crucial as one of the approaches in addressing this complication.

Polyaniline (PANI) is a conductive polymer that also possesses an antibacterial effect [4]. It receives great attention due to its attractive electrical properties through its modification in oxidation and protonation states [5,6]. Polyaniline exists in different forms such as fully oxidized pernigraniline

[a] Corresponding author: syafiqahsaidin@biomedical.utm.my

base, half-oxidized emeraldine base and fully reduced leucoemeraldine [7]. The antibacterial properties of PANIs [4], PANI doped with natural acid [8] and PANI/CZO nanocomposites [9] have been explored. The antibacterial activities of different forms of PANI against four gram positive, five gram negative bacteria and fungus *Candida albicans* were also evaluated by Dhivya *et al.* has shown good antibacterial efficacies [10]. The mechanism of antibacterial effects of PANI also been discussed [10].

Poly (lactic-*co*-glycolic acid) (PLGA) is another polymer which is widely used in implant coating material that effective in controlling material/drug degradation [11]. It has been approved by Food and Drug Administration (FDA) to be restored in human [12]. In this study, PLGA was used to encapsulate ginseng where ginseng has abundant therapeutic benefit such as diabetic inhibitory [13], antioxidant activities [14] and anti-inflammatory effects [15]. Incorporation of PANI in the ginseng encapsulated microcapsules will further enhancing the therapeutic benefit of the microcapsules by providing a reservoir for antibacterial release.

Therefore, PANI was incorporated into ginseng encapsulated PLGA microcapsules to be coated on stainless steel 316L (SS316L) using an electro-deposition technique to equip the coating with antibacterial property. The chemical composition and morphology of the samples were characterized principally by using an attenuated total reflectance-Fourier transform infrared spectroscopy (ATR-FTIR) and scanning electron microscopy (SEM), respectively. The antibacterial efficacy of the samples was also evaluated through a bacterial count test.

2 Materials and Methods

2.1 Materials

Poly(lactic-*co*-glycolic acid) with 85:15 lactide/glycolide ratio and inherent viscosity of 0.63 dL/g was supplied by LACTEL Absorbable Polymers, UK. While ginseng extract was purchased from Dalian Hongjiu Biotech, China. Polyaniline (emeraldine base) with a molecular weight of 20,000 and sodium chloride (NaCl) were acquired from Sigma Aldrich, USA. Dichloromethane (DCM), fully hydrolyzed polyvinyl alcohol (PVA) with a molecular weight of 30,000 and 30% hydrogen peroxide (H_2O_2) were obtained from Merck KGaA, Germany. Acetone, hydrofluoric acid (HF), methanol, glycerol, phosphoric acid (H_3PO_4) and nitric acid (HNO_3) were bought from Friendemann Schmidt, Western Australia. A thin foil medical grade stainless steel 316L with 0.25 mm thickness was obtained from Goodfellow Cambridge Limited, Huntingdon, England.

2.2 Microcapsules Fabrication

Ginseng/PANI was encapsulated in PLGA microcapsules through a double emulsion solvent evaporation method [16]. In the first stage, 400 mg PLGA and 1 mg PANI were dissolved in 10 mL of DCM and 30 mg ginseng extract was dissolved in 1 mL of distilled water, separately. In achieving water-in-oil (W/O) emulsion, both suspensions were emulsified with a homogenizer for 6 min at 20,000 rpm in an ice bath. In order to achieve water-in-oil-in-water (W/O/W) emulsion, the resulting W/O emulsified mixture was instilled into 60 mL of 1.25% w/v PVA solution and homogenized for 10 min at 20,000 rpm in an ice bath.

The mixture was stirred overnight at room temperature to evaporate the DCM. The mixture was then centrifuged at 10,000 rpm for 30 min at 4°C. The centrifugation cause noticeable appearance of microcapsules. The microcapsules were washed once with 0.9% NaCl and twice with distilled water. Finally, the ginseng/PANI encapsulated PLGA microcapsules were stored at 5°C prior to use. The processes were repeated without the addition of PANI to fabricate ginseng encapsulated PLGA microcapsules.

2.3 Sample Preparation

2.3.1 Pre-treatment

Medical grade stainless steel 316L foil was cut into square shape of 10 mm × 10 mm and undergoes several pre-treatment processes [17] included ultrasonic cleaning, electro-polishing and acid-dipping. The metals were each ultrasonic cleaned in acetone, DI water and methanol for 10 min. Then, the cleaned metals were electro-polished for 3 min at 1.5 A in an electrolyte solution (50% glycerol, 15% DI water and 35% H_3PO_4). Acid-dipping was done by dipping the electro-polished metals in an acid solution (10% HNO_3, 2% HF and 88% DI water) for 30 sec to remove the remaining electro-polishing ions that adhered on the metal surface. The pre-treated metals were then rinsed with DI water, followed by drying using an air compressor.

2.3.2 Microcapsules Deposition

The ginseng/PANI encapsulated PLGA microcapsules and the ginseng encapsulated PLGA microcapsules, denoted by PLGA/g/PN and PLGA/g, respectively, were deposited on the pre-treated metals using an electrodeposition technique following the procedures by Wang *et al.* [18] with some alterations. An electrolyte solution was prepared by adding 1% w/v microcapsules into DI water. The conductivity of the electrolyte was improved with the addition of 0.15 M NaCl and 100 mM H_2O_2. The electrolyte was stirred for 30 min at room temperature. The microcapsules were electro-deposited using a direct current (DC) power supply (72-8690A, TENMA, Japan) in a laminar flow. The pre-treated metal was clasped at the cathode while a reference electrode was clasped at the anode. Both cathode and anode were then immersed in an electrolyte solution at room temperature. The electrodeposition was conducted at 2 mA current density for 40 sec. Finally, the coated metals were washed with DI water and dried in a vacuum desiccator prior to analyses.

2.4 Characterization analyses

2.4.1 Chemical Composition Analysis

The PLGA/g/PN, PLGA/g and pre-treated metal were characterized using attenuated total reflectance-Fourier transform infrared spectroscopy (ATR-FTIR). The surface composition of the samples was determined by ATR-FTIR (Nicoler iD5, Thermo Scientific, USA) using ZnSe crystal at 4 cm^{-1} resolution. The spectra were recorded within 500 - 4000 cm^{-1} frequency range and collected at 32 average scans.

2.4.2 Morphological Analysis

The microstructure and morphology of PLGA/g/PN, PLGA/g and pre-treated metal were visualized using a tabletop SEM (Hitachi, TM300, Japan), viewing at an accelerating voltage of 15 kV, under magnification of 1000× and 3000×. A thin layer of gold was sputtered on the microcapsule coating to avoid charging under electron beam, prior to the SEM analysis.

2.5 Antibacterial analysis

The antibacterial activity of the samples was evaluated with Gram-negative *Eschericia. coli* (*E. coli*, ATCC 25922) through a quantitative analysis. The pre-treated metal, PLGA/g and PLGA/g/PN were sterilized with ultraviolet (UV) light on each surface for 30 min. The samples were then immersed in 5 mL of bacterial suspension that contained $1x10^6$ cells/mL bacteria. All inoculated samples were incubated for 3 h in a shaking incubator at 200 rpm. After incubation, the samples were removed and rinsed two times with phosphate buffer saline (PBS) to remove non-attached bacteria. The samples were placed in a new sterile centrifuge that contained 5 mL of Luria-Bertani (LB) broth. The adhered bacteria on the sample surfaces were detached using a vortex for 30 sec. The bacterial suspensions were diluted to six dilution series. A drop plate method was then used to grow the bacteria on LB agar. The method was performed by pipetting 10 μL of each dilution on LB agar to be incubated for 24 h at 37°C. The colony-forming units (CFUs) were counted manually to obtain the number of adherent bacteria. Equation 1 was finally used to calculate the bactericidal ratio.

$$Bactericidal\ ratio = \frac{(CFU_{control} - CFU_{experiment})}{CFU_{control}} \times 100 \quad (1)$$

3 Results and Discussion

3.1 Characterization analyses

3.1.1 Chemical Composition Analysis

Figure 1 shows the ATR-FTIR spectra of pre-treated metal, PLGA/g and PLGA/g/PN. No identical peak was depicted on the pre-treated metal, defining the pre-treated metal surface was clear from apparent functional group. The presence of PLGA and ginseng were validated on the PLGA/g/PN and PLGA/g when C–O peaks were noticed at 1089 and 1182 cm^{-1} due to the existence of C–O functional groups. However, another identical peak of PLGA and ginseng was only noticed on the PLGA/g/PN, denoted by C–H peak at 1459 cm^{-1}.

Strong peak of carbonyl C=O stretching vibration at 1741 cm^{-1} was also recorded only on the PLGA/g/PN, demonstrating the composition of PLGA. Meanwhile, the composition of ginseng was clarified by the appearance of strong and broad peak of O–H band (3000 – 3500 cm^{-1}) on both PLGA/g/PN and PLGA/g. A medium peak of C=C stretch alkene was also observed only on the PLGA/g/PN at 1644 cm^{-1}, presenting the functional group of ginseng. However, the presence of PANI on the PLGA/g/PN was not identified clearly as some peaks such as N–H, C–N and C=C from the PANI emeraldine compound [10,19] might overlay with other peaks arising from the PLGA and ginseng. From the results of ATR-FTIR, the ginseng/PANI encapsulated PLGA microcapsules were intensely deposited on the pre-treated metal with the facilitation of PANI due to the appearance of strong PLGA and ginseng functional groups on the metal surface.

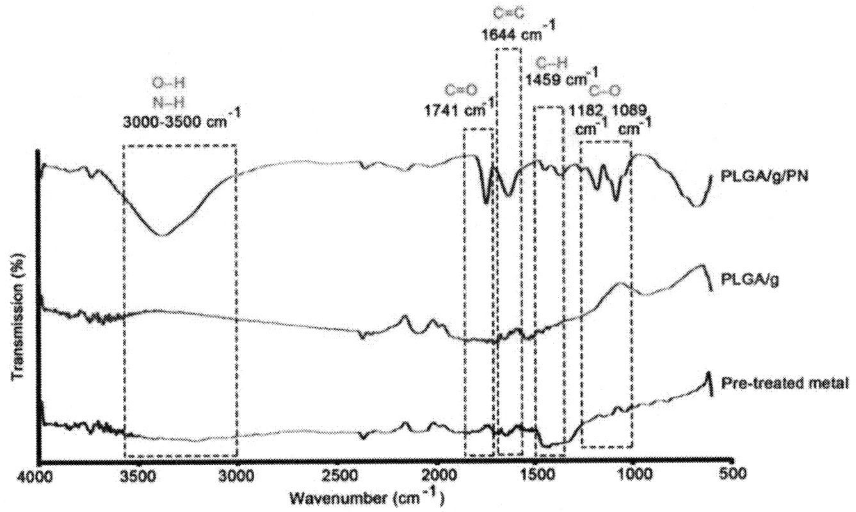

Figure 1. ATR-FTIR spectra of pre-treated metal, PLGA/g and PLGA/g/PN.

Figure 2 shows the morphology of pre-treated metal, PLGA/g and PLGA/g/PN. A clear and smooth surface was observed on the pre-treated metal due to the pre-treatment processes of ultrasonic cleaning, electropolishing and etching. The passive layers of stainless steel 316L were dissolved during the electropolishing caused the removal of micro-scratch and flaws. While the etching was intended to remove the phosphate layer adhered on the metal surfaces [20]. There was a few formation of microcapsules on the PLGA/g where greater volume of microcapsules was visualized on the PLGA/g/PN. The incorporation of PANI played a role in the attachment of microcapsules on the pre-treated metal where this material acted as a conductive agent [21] in driving the movement of microcapsules towards the metal surfaces during the electro-deposition process.

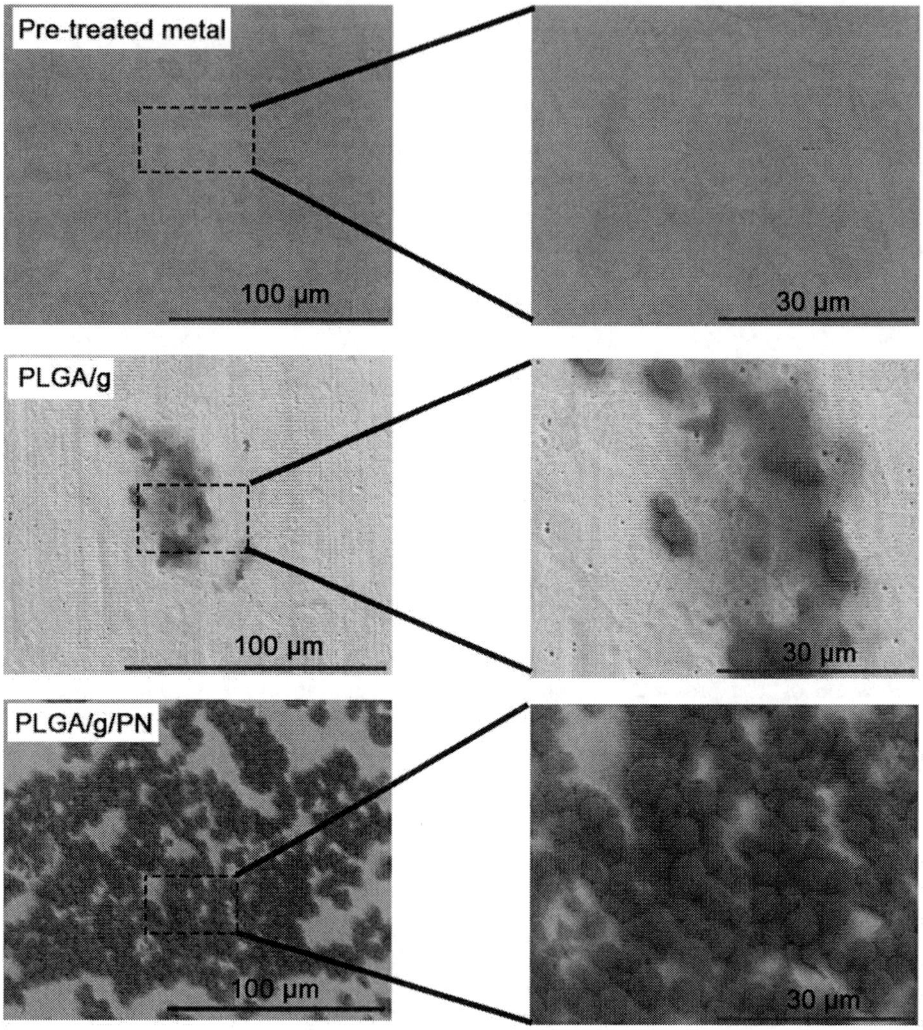

Figure 2. SEM images of pre-treated metal, PLGA/g and PLGA/g/PN.

3.2 Antibacterial Analysis

The data of antibacterial quantitative analysis were listed in Table 1 while the antibacterial activity of the samples on gram negative bacteria, *E. coli* was shown in Figure 3. These results were based on the bacteria adhered on the metal surfaces. Evidently, it was indicated that the pre-treated metal has no antibacterial effect. The bactericidal ratio of the PLGA/g/PN (87.64%) against *E. coli* was higher than the PLGA/g (37.45%). The presence of PANI in the PLGA/g/PN killed the adherent *E. coli* even though the surfaces might have more bacterial attachment due to the rough structure. A study by Nanlin *et al.* [4] stated that the bacterial death occurred due to acidic dopants on the molecular chain of PANI that reacted with the bacteria. Besides, the electrostatic of PANI molecules also played a role in providing the antibacterial effect [4].

Table 1. CFU/mL and bactericidal ratio of *E. coli* on pre-treated metal, PLGA/g and PLGA/g/PN

Sample	**CFU (1x10^6 cells/mL)**	**Bactericidal ratio (%)**
Pre-treated metal	2.67 ± 1.53	0.00
PLGA/g	1.67 ± 0.57	37.45
PLGA/g/PN	0.33 ± 0.57	87.64

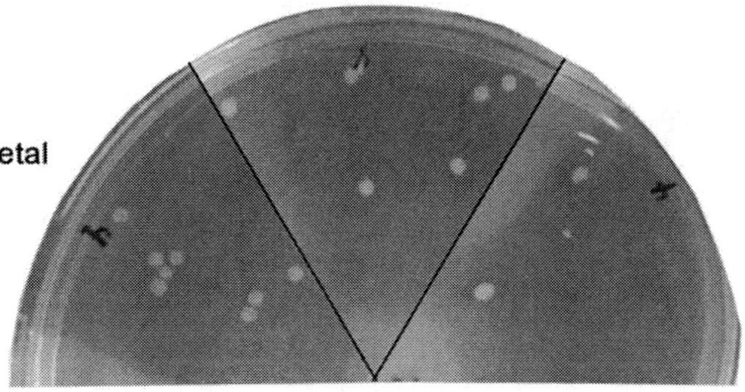

Figure 3. The antibacterial activity of the samples on Gram-negative *Eschericia. Coli.*

4 Conclusion

The incorporation of PANI in the ginseng encapsulated PLGA microcapsules has served the microcapsules with antibacterial effects. The microcapsules were coated onto stainless steel 316L using electrodeposition technique. The formation of microcapsules was proved through ATR-FTIR and SEM images, where the microcapsules successfully deposited onto metal surface. It also provides antibacterial effects that may be usable in metallic implants when the bactericidal ratio on microcapsules incorporated PANI shows higher than without the addition of PANI. Therefore, metallic implant equipped with PANI will reduce the probability of infection and inhibit the bacterial growth while inducing the formation of ginseng encapsulated PLGA microcapsules.

5 Conflict of Interest

There is no conflict of interest.

Acknowledgement

This study was supported by Malaysian Ministry of Higher Education through a Research University Grant (RUG) Tier 1 [Q.J130000.2545.18H40]. The authors would like to appreciate Tan Sri Ainuddin Wahid scholarship and thank Mohamad Amin Jumat for his contribution in conducting antibacterial analysis.

References

1. Y. Yu, G. Jin, Y. Xue, D. Wang, X. Liu, J. Sun, Acta. Biomater. **49**, 590-603 (2017)
2. R. Martin, E. Paul, K. E. P. G. Richard, F. Reno, W. Maja, Pacing Clin. Electrophysiol. **38**, 247-253 (2015)
3. C.W. Chan, L. Carson, G. C. Smith, A. Morelli, S. Lee, Appl. Surf. Sci. **404**, 67-81 (2017)
4. N. Shi, X. Guo, H. Jing, J. Gong, C. Sun, K. Yang, J. Mater. Sci. Technol. **22**, 289-290 (2006)
5. A. Mostafaei, A. Zolriasatein, Pro. Nat. Sci-Mater. **22**, 273-280 (2012)
6. S.I. Abd Razak, I.F. Wahab, F. Fadil, F.N. Dahli, A.Z. Md Khudzari, H. Adeli, Adv. Mater. Sci. Eng. **2015**, 1-19 (2015)
7. E. Song, J.W. Choi, Nanomaterials **3**, 498-523 (2013)
8. J.L. Mu, W.J. Fan, S.Y. Shan, T.W. Hu, Y.M. Wang, Q.M. Jia, Adv. Mater. Res. **650**, 249-252 (2013)
9. X. Liang, M. Sun, L. Li, R. Qiao, K. Chen, Q. Xiao, et al. Dalton Trans. **41**, 2804-2811 (2012)
10. C. Dhivya, S.A.A. Vandarkuzhali, N. Radha, Arab. J. Chem. (2015)
11. D.J. Hines, D.L Kaplan, Crit. Rev. Ther. Drug Carrier Syst. **30**, 257-276 (2013)
12. T. Ponnusamy, L.B. Lawson, L.C. Freytag, D.A. Blake, R.S. Ayyala, V.T. John, Biomatter **2**, 77-86 (2012)
13. S.S. Shin, M. Yoon, J. Ethnopharmacol. **210**, 80-87 (2018)
14. T. Ramesh, S.W. Kim, S.Y. Hwang, S.H. Sohn, S.K. Yoo, S.K. Kim, Nutr. Res. **32**, 718-726 (2012)
15. S.Y. Lee, J.J. Jeong, S.H. Eun, D.H. Kim, Eur. J. Pharmacol. **762**, 333-343 (2015)
16. S.K. Lukman, S. Saidin, Z.K. Lo, R.H. Al-Ashwal, Paper presented at the Proceedings of the 2017 International Conference on Computational Biology and Bioinformatics 101-104 (2017)
17. F. Lewis, M. Cloutier, P. Chevallier, S. Turgeon, J.J. Pireaux, M. Tatoulian, et. Al, ACS Appl. Mater. Interfaces, **3**, 2323-2331 (2011)
18. Y. Wang, X. Guo, R. Pan, D. Han, T. Chen, Z. Geng, et al, Mater. Sci. Eng. C **53**, 222-228 (2015)
19. N.P. Chauhan, R. Ameta, R. Ameta, S.C. Ameta, Indian J. Chem. Technol. **18**, 118-122 (2011)
20. M. Haïdopoulos, S. Turgeon, C. Sarra-Bournet, G. Laroche, D. Mantovani, J. Mater. Sci. Mater. Med. **17**, 647-657 (2006)
21. A. Sarvi, A.B. Silva, R.E. Bretas, U. Sundararaj, Polym. Int. **64**, 1262-1267 (2015)

A Copper Oxide (CuO) Thin Films Deposited by Spray Pyrolysis Method

Youcef Bellal[1,2,a], Antar Bouhank[1], Hacene Serrar[1], Tunç Tüken[3] and Gökmen Sığırcık[3]

[1]*Thin films development and applications unit Setif, Research Center in Industrial Technologies (CRTI)P.O.Box 64,Cheraga 16014Algiers, Algeria*
[2]*Faculty of Technology, Department of Engineering Process, Laboratory of Electrochemistry, Molecular Engineering and Redox Catalysis (LEIMCR), Ferhat Abbas University Setif-1, Algeria*
[3]*Faculty of Science and Letters, Çukurova University, Adana, Turkey*

Abstract. A simple and low-cost procedure (spray pyrolysis) was used to elaborate a copper oxide thin films on ordinary glass substrates. A copper nitrate was used and dissolved in two different solutions (Water, Methanol) S_1 and S_2 respectively in order to obtain an equal concentration; $C_{S1,S2}$=0.5M. The spray pyrolysis deposition made at fixed temperature T=500°C and different volumes of S_1 or S_2 on the glass substrates. The X-ray diffraction (XRD), scanning electron microscopy (SEM) and UV-vis spectrophotometry were used to determinate the structural, morphological and optical properties of CuO thin films. The X-ray diffraction patterns confirm the presence of the polycrystalline phase of CuO as monoclinic crystal structure with preferential orientation along (110), (002), (111), (200) and (020). Their optical band gaps ranged from 3.95 to 4.02eV for thin films made with S_1, and from 1.6 to 1.95eV for thin films made with S_2 with a high absorbency in the visible region, which is in agreement with the values of the literature.

1 Introduction

The waters pollution is an undesirable phenomenon caused by the technological development of the human being during these last centuries until the next day and which menaced the living beings to survive in this planet earth. For this, physical and / or chemical techniques are developed to fight against this serious problem [1]. In our research case we are interested in Photocatalysis, which is a simple and less expensive procedure based on the principle of activating a semiconductor using the energy provided by light. The semiconductor is considered a catalyst. Its principle is close to the heterogeneous catalysis where the oxidation-reduction reaction takes place on the surface of the catalyst [2], based on the absorption of photons with energy greater than or equal to the band gap of the photocatalyst to generate electrons in the conduction band and holes in the valence band. The electron-hole pairs thus generated can recombine or react with organic compounds adsorbed on the surface of the semiconductor via the oxidation-reduction processes. In aqueous media, the holes trap water molecules that are transformed into $HO^{\bullet}$ hydroxyl radicals. These hydroxyl radicals react with

[a] Corresponding author : y.bellal@crti.dz

organic compounds in solution (photo indirect degradation). These same holes can directly oxidize organic molecules (direct photo degradation).

Photocatalysis is generated in the 1970s. This technique consists of spontaneous irradiation with daylight or an artificial lamp with a semiconductor [3]. It is considered a new active and cost-effective abatement technology for the treatment of residual organic substances and the modification of hazardous chemicals in different forms [4]. The semiconductor material undergoes radiation excitation which allows an electron of the valence band to be ejected into the conduction band [3].

In the previous years the decontamination of the undesirable (toxic) organic or mineral products found in the water is generally done by semiconductors in the form of fine particles immersed in suspensions, the regeneration and recovery of which require delicate techniques and expensive (membranes, separation column, etc. ...).

Today researchers are interested in the technique of thin films for the application of photocatalytic which involves the fixing of a thin film of a material on another material, called "substrate" whose purpose of give particular properties to the surface of the room [5]. The elaborate thin layers are generally influenced by the methods and conditions of the preparation such as the deposition technique (physical or chemical), the annealing temperature, types of substrates etc. [6]. As an example of the ultrasonic spray technique of ZnO films, the optoelectronic properties are largely influenced by the thickness of the film and the temperature of the substrate [7]. Above all, for a polycrystalline film, microscopic limit pores that depend on film thickness control the properties of photoconduction [7]. For a film, the thickness can be varied by the deposition time and also by the concentration of the solution. Also the temperature of the substrate that has an effect on the layer during growth.

2 Materials and Methods

We used in our work a simple means of deposition, it is an empty perfume spray bottle of 15 ml volume filled with a desired precursor solution S_1 or S_2, a hot plate at high temperature (550°C), ultra sonic apparatus, ordinary glass substrates and solvents for cleaning and preparation of S_1 and S_2 solutions. the precursor (copper nitrate) and the solvents are Aldrich brand.

- The solutions S_1 and S_2 are prepared at room temperature under the stirring effect for one hour in the open air.

- Surface treatment of the substrates using a mixture of equimolar solvent (acetone / methanol) in an ultra sonic apparatus at T=40°C for 10 minutes.

3 Results and Discussion

3.1 Optical Characterization

It is well known that thin layers of CuO have a p-type conductivity and are the direct result of Cu gaps in the assembly of the structure which leads to the formation of holes in the valence bond [8]. allowed us to use equation (1) and (2) for the calculation of gap energy.

Fig.1. shows the optical absorbance spectra of thin films deposited at different solvents. As expected, the resulting CuO thin films exhibit: low and strong absorption in the visible region for thin films made with S_1 and S_2 respectively; however, it declines after 800nm to the extent that the transparency is considerable from 900nm to 1100nm. These are the basic characteristics of a solar selective absorber [9].

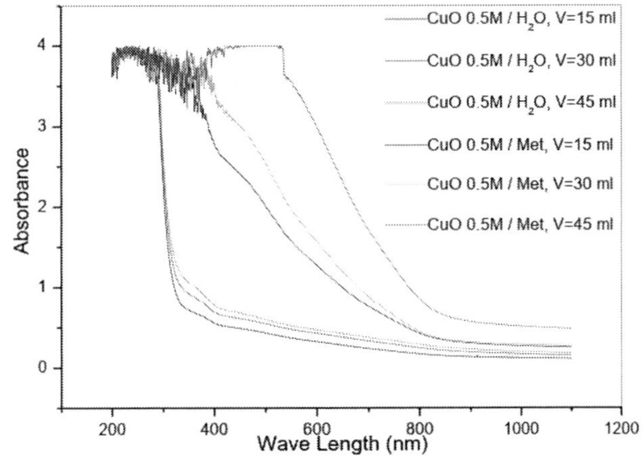

Figure. 1. Optical Absorbance of CuO Thin Films Deposed at Different volumes (S_1 or S_2) at T=500°C

In order to appreciate the gap energy of CuO thin films, first, the experimental data were extracted and reformed to absorption coefficient (α) via equation (1) [10.11] as follows:

$$\alpha = \frac{1}{t}\, ln[\frac{(1-R^2)}{2T} + \sqrt{\frac{(1-R)^4}{4T^2}} + R^2] \quad (1)$$

knowing that:
t: Thickness
R: Reflectance
T: Transmittance

Taking into account that the thickness *t* is of the order nano meter, the gap energy Eg and the absorption coefficient α can be calculated according to the relation of Tauc (2).

$$\alpha h\upsilon = A(h\upsilon - E_g)^n \quad (2)$$

Where *h* is the photons energy , *A* is an independent constant of the energy, and n is ½ for the direct allowed transition. To this, the gap energy can be estimated by plotting $(\alpha h\nu)^2$ with respect to $h\nu$ one extrapolating the linear part of the graph at $(\alpha h\nu)^2$=0 Fig. 2 and 3 [12], whose resulting band gap values are in good agreement with the literature [13-16]. The wavelengths of the corresponding thin layers and the gap energies are mentioned in Table 1 using equation (3).

$$E_g = hc/\lambda \approx 1240/\lambda \quad (3)$$

Where:
c: Speed of light

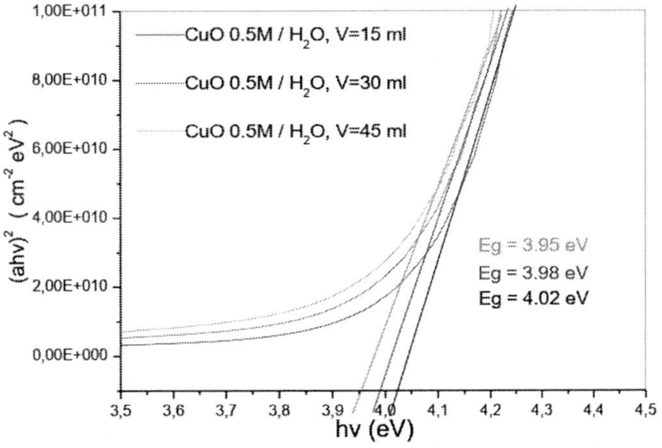

Figure. 2. Gap energy of CuO thin films produced at different volumes from S_1 at T=500°C.

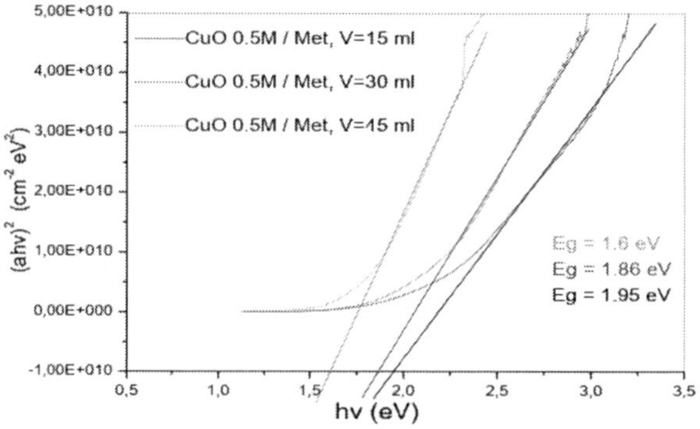

Figiure. 3. Gap energy of CuO thin films produced at different volumes from S_2 at T=500°C.

Table 1. The wavelength values correspond to the gap energies of the thin films elaborated by different volumes of S_1 and S_2 at T=500°C.

Sample	Energy gap (eV)	Wavelength (nm)
CuO 0.5M/H_2O,V=15ml	3.95	313.92
CuO 0.5M/H_2O,V=30ml	3.98	311.56
CuO 0.5M/H_2O,V=45ml	4.02	308.46
CuO 0.5M/Met,V=15ml	1.6	775.00
CuO 0.5M/Met,V=30ml	1.86	666.66
CuO 0.5M/Met,V=45ml	1.95	635.89

It seems that among the factors that can possibly modify the gap energy is the size of the crystallites of the CuO thin films which are mainly responsible for the band gap variation. In this study, the correlation between crystallite size and band gap was discussed earlier [17,18]. As a result, by decreasing the size of the crystallites by changing the polarity of the solvent towards the less polar, the band gap has generally increased.

3.2 Structural Characterization

3.2.1 XRD Characterization

The X-ray diffraction diagrams (fig. 4) of the CuO thin films produced at different volumes of S_1 or S_2 at T= 500°C. show a high intensity of peaks observed at 2θ = 35.55 ° and 38.73 ° to the diffraction planes (002) and (111) and some small peaks assigned to (110), (200), (-202), (020), (202), (-113), (-311), (220)) and (311) observed at 2θ = 32.52.41.98, 48.74, 53.68, 58.30, 61.56, 66.25, 68.13 and 72.34 °, respectively. It can be noticed that the XRD model has not changed much by the precursor solvent change. All the apparent atomic planes show the formation of the CuO phase with monoclinic crystalline structure. The diffraction peaks (002) and (111) of CuO were high in the process of increasing the thickness of the thin layer due to the increase in precursor concentration [19].

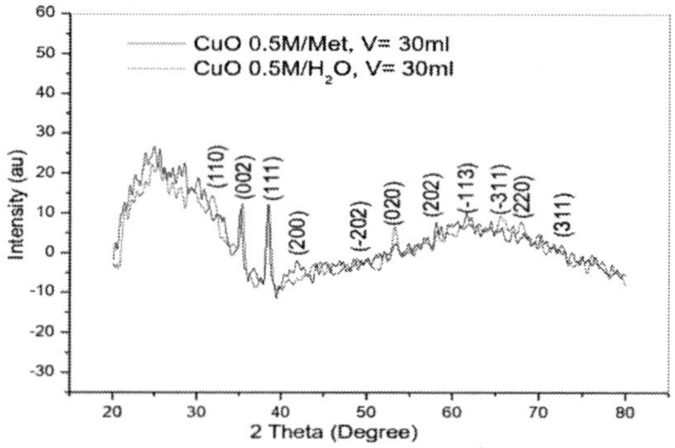

Figure. 4. X-ray diffraction of CuO thin films made at different solvents S_1 or S_2 at V=30ml and T=500°C.

3.2.2 SEM Characterization

Two surface morphologies of the thin films produced with different solvents (water or methanol) were observed. Spherical grains tightly packaged with different sizes from 40 to 160 nm in diameter for thin films made with S_1 [20]. Well-bonded and overlapped coral-shaped grains are observed for thin layers made with S_2 (Fig. 5).

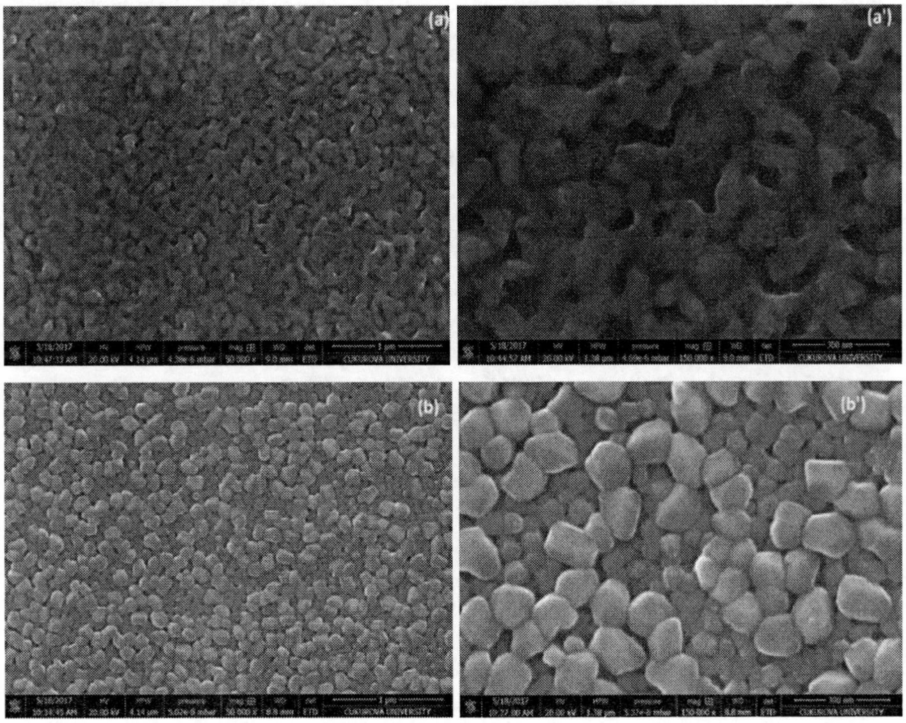

Figure. 5. SEM images at T=500°C, V=30ml: a, a '(CuO/Methanol), b, b' (CuO/H_2O).

4 Conclusions

The thin films obtained by the spray pyrolysis method using the copper nitrate precursor us to conclude the following results:

- ➢The speed and the affordable cost of the technique.
- ➢The UV-Visible characterization showed that all films exhibit ultraviolet optical transmission and their optical band gaps ranged from 3.95 to 4.02eV for thin films made with S_1, and from 1.6 to 1.95eV for thin films made with S_2.
- ➢The X-ray diffraction patterns confirm the presence of the polycrystalline phase of CuO as monoclinic crystal structure.
- ➢SEM images show that a material (CuO) settles differently on the surface of substrates by noticing a homogeneous roughness and nanoscale grains with high wettability for CuO deposited with S_2 unlike S_1. so this roughness is a very positive factor for the photo-catalytic degradation phenomenon.

Acknowledgements

The authors gratefully acknowledge the financial support from the Algerian Ministry of Higher Education and Scientific Research. The authors would like to thank Professor

Abderrahmane Bouabellou, Thin Films and Interfaces Laboratory, Frere Mentouri University, Constantine, Algeria, for helpful.

References

1. V. Kazpard, Z. Saad, A.-G. El Samrani et K. Slim. European journal of water quality - Journal européen d'hydrologie, 10, p. 71-77, 2005.
2. Vincent Felisab. Claude De Bellefonb. Pierre Fouillouxb. Daniel Schweichb. Applied Catalysis B: Environmental..20. p. 91-100, 1999.
3. Atef A., Elaboration De TiO2 Sous Forme De Couche Mince Dopée Et Nanotubulaire : Caractérisation Électrochimique Et Performance Photocatalytique. Thése de Doctorat, Université El Manar,Tunisie, 132, 2013
4. Ahmed S., Rasul M., Martens W.N., Richard B., Hashib M. A., Advances in Heterogeneous Photocatalytic
   Degradation of Phenols and Dyes in Wastewater: A Review, 27, 2010.
5. Xingwang Z, Minghua Z, Lecheng L, Mater. Chem. Phys. p. 73-79, 2005.
6. K. Zine Ali, Thèse de magister, Université Houari Boumediene Alger, 2003.
7. Ayadi Meriem, Thèse de magister Université de Skikda, 2006.
8. Jeong YK, Choi GM. Journal of Physics and Chemistry of Solids. 57(1), p. 81-84, 1996.
9. Shabu R, Raj AME, Sanjeeviraja C, Ravidhas C..Materials Research Bulletin. 68:18, 2015.
10. Yahia IS, Farag AAM, El-Faify S, Yakuphanoglu F, Al-GhamdiAA.. International Journal for Light and Electron Optics.127(3), p. 1429-1433, 2016.
11. Hassanien AS, Akl AA.. Superlattices and Microstructures.89, p. 153-169, 2016.
12. Tauc J, Grigorovici R, Vancu A..physica status solidi (b).15(2), p. 627-637, 1966.
13. Saravanan V, Shankar P, Mani GK, Rayappan JBB.. Journal of Analytical and Applied Pyrolysis.111, p. 272-277, 2015.
14. Kose S, Atay F, Bilgin V, Akyuz I. Materials Chemistry and Physics. 111(2-3), p. 351-358, 2008.
15. Mehdi Dhaouadi et al., American Journal of Physics and Applications, 6(2),p. 43-50, 2018.
16. M. Lamri Zeggar et al., International Journal of Physical and Mathematical Sciences, 9(10), p. 632-635, 2015.
17. Akgul FA, Akgul G, Yildirim N, Unalan HE, Turan R.. Materials Chemistry and Physics., 147(3), p. 987-995, 2014.
18. Gupta Y, Arun P, Naudi AA, Walz MV, Albanesi EA, Thin Solid Films, 612, p. 310-316, 2016.
19. R. Shabu, A. Moses Ezhil Raj, C. Sanjeeviraja, C. Ravidhas, Materials Research, Bulletin, 68, p. 1–8, 2015 .
20. P.Venkateswari,P. Thirunavukkarasua, M. Ramamurthy, M. Balaji, J. Chandrasekaran, International Journal for Light and Electron Optics, S0030-4026(17)30434-5, 2017.

Additive Manufacturing in Vascular Stent Fabrication

Yang Lei[1, 2, #], Xin Chen[1, 2, 3, 4 #], Lei Zhang[1, 2, a], Lei Li[5], Shuangzhu Kang[1, 2], Chengjin Wang[1, 2] and Wei Sun[1, 2, 6]

[1] *Department of Mechanical Engineering, Tsinghua University, 100084, China*
[2] *111 'Biomanufacturing and Engineering Living Systems' Innovation International Talents Base, 100084, Beijing, China*
[3] *Ministry of Industry and Information Technology Equipment Industry Development Center, Beijing, 100846, China*
[4] *China Center for Information Industry Development, Beijing, 100048, China*
[5] *Beijing Huaxin Hospital First Hospital of Tsinghua University, Beijing, 100016, China*
[6] *Department of Mechanical Engineering, Drexel University, Philadelphia, PA, United States of America*
[#] *These authors contributed equally to this work.*

Abstract. High-efficiency formation of personalized stent by additive manufacturing (3D printing) has gained deal of attention and research in interventional and personalized medicine. In this article, the structural characteristics of vascular scaffolds and the application and innovation of additive manufacturing technology in the process of angioplasty are reviewed. In the future, with the continuous maturity of additive manufacturing technology, it is expected to be an important part of interventional precision medicine to manufacture personalized vascular stent.

1. Introduction

Vascular trauma and defect, as well as cardiovascular and cerebrovascular diseases, have become the main threats to human life and health. Due to the good performance on reconstruction of vascular structure, stents implantation has become the main treatment for vascular diseases[1]. Based on different fabricate structures, vascular stents can be divided into dense mesh stents and stent grafts. Dense mesh stents[2] include bare metal stents (BMS), drug eluting stents (DES) and bioresorbable stents (BRS)[3, 4]. After expanded, such stents can support blood vessels, while be fixed by the pressure produced by the vessel wall, in which case, the lumen unobstructed will be guaranteed. Dense mesh stents are used to treat coronary heart disease, ischemic stroke, peripheral vascular stenosis, etc.[5] Stent graft is a type of vascular stent with a fabric coat, it is expandable like bare metal stents, and the film creates a contained tube to replace the vascular wall. It is mainly used to repair aortic dissection, true aortic aneurysm, pseudoaneurysm, aortic penetrating ulcer and other diseases. It is also used in the treatment of intracranial aneurysms[6].

The current manufacturing methods of vascular stents are standardized, that is, the diameter and length of stents are serialized or standardized. However, there are individual differences in the diameter and shape of real blood vessels in the human body, which makes standardized stents difficult to match the personalized needs. If the stent size is not consistent with the patient's real vascular size after implantation, the therapeutic effect will be affected. Too small diameter of the stent will lead to poor

[a] Corresponding author : stoneszhang@mail.tsinghua.edu.cn

adherence to the wall of the stent, resulting in internal leakage or displacement. Too large diameter can rupture blood vessels, leading to treatment failure. At the same time, the shape of real blood vessels in the human body is curved, therefore, some standardized stents may "shrivelled" after implantation, because they cannot adapt to the physiological structure of blood vessels. Therefore, the individualized customization of vascular stent is required in clinic [7-10].

Additive manufacturing (3D printing) can not only achieve the personalized vascular stent in a short time, but also avoid the treatment failure caused by the poor compliance of the traditional stent. In addition, compared with other personalized fabricate technology, the 3D printing efficiency of vascular stents is higher. Now, with the maturity of the additive manufacture technology, it has gained widely concern in the field of vascular stent fabrication and shown a good application prospect in the clinical treatment.

In this paper, the structural characteristics of vascular scaffolds are reviewed, as well as the status and prospect of the additive manufacture technology in vascular stent fabrication.

2. Dense mesh stents

2.1 Structural features of dense mesh stents

The dense mesh stent is an expanded mesh structure that provides sustained support in order to open a narrow occlusion vessel. It is mainly used for the treatment of cardiovascular and cerebrovascular diseases that are serious threats to human health such as coronary heart disease and stroke. With the gradual improvement of product performance, endovascular stents have undergone three development stages: BMSs, DESs and BRSs[11].

2.1.1 Bare metal stents

In 1964, Dotter et al. first proposed the concept of endovascular stent. In 1987, Swiss doctor Ulrich Sigwart[12] first implanted a metal stent into a coronary artery to support a stenotic obstruction vessel, so that the stent can reduce the elastic retraction of the blood vessel in order to cure a coronary heart disease. Since then, endovascular stents have been widely used in interventional therapy. The materials of BMSs is generally medical 316L stainless steel, titanium alloy, cobalt alloy, tantalum metal, etc., which have strong supporting force[13]. However, the traditional permanent metal stent will inevitably cause immunological rejection after implantation, which is easy to cause intimal hyperplasia, stimulate the proliferation of smooth muscle cells and cause postoperative restenosis, which seriously affects the long-term patency rate of the vascular stent[14]. Clinical statistics show that after bare metal stent implantation, the rate of restenosis reaches 20%.

2.1.2 Drug eluting stents

DES is developed based on BMS. DES is mainly composed of a metal stent, a drug carrier and a drug coating. DES can inhibit the proliferation of vascular endothelial cells, the aggregation of plasma proteins and white blood cells, and prevent the restenosis of blood vessels through slowly release of drugs in the blood vessels[15]. DES has been developed by two generations. The first generation is coated with rapamycin or paclitaxel on a stent of medical stainless steel. The most representative products are *Cypher* (Figure 1.A) by *Johnson&Johnson*[16] and *Taxus* (Figure 1.B) by *Boston Scientific*[17]. Since these drugs are prone to leukocytes and coagulation factors attached to the scaffold, the incidence of late stent thrombosis in the first-generation DES is significantly higher than that in BMS[18]. The second-generation DES is represented by the *Abbott XienceV* (Figure 1.C)[19], using cobalt-chromium alloy carry zotarolimus or eromox. The rate of restenosis is almost the same as that of the first-generation DES, but the incidence of late-stage stent thrombosis is significantly reduced[20].

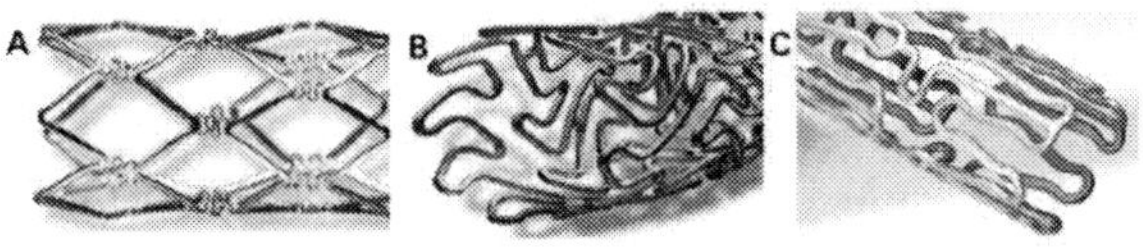

Figure 1. 1st & 2ed generation DESs. (A) Cypher by Johnson&Johnson; (B) Taxus by Boston Scientific; (C) XienceV by Abbott

2.1.3 Bioresorbable stent

The clinical data shows that the revascularization can be completed within 6-9 months after the implantation. At that time, as a foreign substance in vivo, permanent stent will hinder the recovery of vascular function, so the bioresorbable stent becomes a new research hotspot. BRSs can be divided into metal stents and polymer stents. The materials of the BRSs are mainly iron-based alloys, magnesium-based alloys and zinc-based alloys[21]. These metals are essential trace elements in human body and have good biocompatibility. However, how to match the degradation rate and the rate of blood vessel regeneration is a problem that metal bioresorbable stents need to solve. The polymer scaffold is generally made of a mixture or copolymer of poly-L-lactide (PLLA), polyglycolic acid (PGA), and polycaprolactone (PCL) [22-28]. In 2006, Abbott obtained the Guidant DES line from Boston Scientific for $6.4 billion. Based on the Guidant BVS, polylactic acid BRS from Abbott obtained CE certification. In 2016, the Abbott Absorb (Figure 2)[29] was approved by the FDA for interventional treatment of coronary artery disease, which is the first degradable vascular stent in clinical use in the United States.

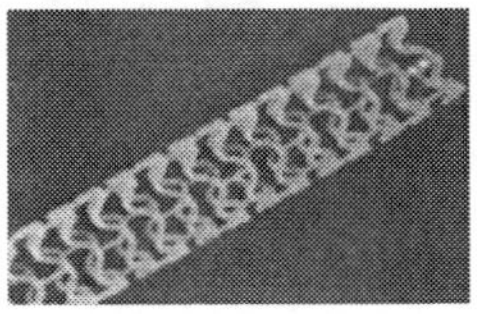

Figure 2. Abbott Absorb BRS.

2.2 Additive manufacturing of dense mesh stents

In the research of vascular scaffold materials, the molding process has experienced the stages of filament winding, photochemical etching, injection molding and laser cutting[30]. In the early stage, stainless steel stents were fabricated by wire winding, and each unit was connected by laser local welding. However, the small size and complex structure of vascular stent result in inability to locate at the weld. The photochemical etching process avoids the positioning process, it uses ultraviolet light to etch the plane model of the vascular stent on the stainless steel sheet, which is then folded and welded into shape. However, the mesh structure processed by this technology is relatively soft, and it is easy to break in the process of crimping forming, while the welding points are arranged on one side of the stent, which affects the mechanical properties[31]. Injection molding preparation vascular stent process does not have above problems, but the vascular stent has high requirements for the injection molds. Moreover, the molding unevenness during the injection molding process is serious. Laser cutting technology has become a commonly used vascular stent processing technology because of its high processing precision and no change in material physicochemical properties[32]. However, toxic gases are generated during laser engraving[33], and the stent support section is rectangular, which may cause local eddy currents in the blood, which induce leukocyte aggregation and lead to restenosis[34].

These years, additive manufacturing technology has been concerned in vascular stent manufacturing. Since the formation of metal materials requires consideration of oxidation and other issues, the current additive manufacturing vascular stents are concentrated in the bioresorbable polymer stents.

In 2013, Flege et al.[35] of Aachen University of Technology, Germany used PLLA and PCL materials respectively to process BRSs using laser selective melting technology (SLM) for the first time, and pointed out that SLM technology has a good prospect in the manufacture of BRSs. However, the surface accuracy of the stent produced by this method is not high (Figure 3.A), which limits its development. In 2015, Park et al.[36] of the Korea Institute of Machinery and Materials used bio-cutting technology to manufacture a bioresorbable sirolimus-eluting PCL stent, but this technique makes it difficult to personalize the stent. In addition, the stents manufacture process by Flege and Park take more than 10h.

In 2015, Tumbleston et al.[37] of Carbon3D, U.S. developed continuous liquid interface production (CLIP) technology. The principle of CLIP: The UV projection solidifies the photosensitive resin, while the oxygen inhibits the solidification. The liquid resin at the bottom of the tank maintains a stable liquid area due to the contact with oxygen, thus ensuring the continuity of the solidification. This technology not only increases the speed of 3D printing by 25 to 100 times, but also make the layering infinitely delicate, which solves the problem of low precision of product surface. This technology plays a vital role in the rapid prototyping of high quality intravascular stents.

In 2016, van Lith et al.[38] of Northwestern University, U.S. synthesized a new citrate-based degradable polymer material, using micro CLIP technology to create customizable individualized arrow types. This stent (Figure 3.B) has good elasticity, strength, oxidation resistance and biodegradability, and the micro CLIP technology compresses the stent forming time to 70 minutes. According to the radial compression test, the maximum supporting force of 0.1n/mm can be provided when the bracket is compressed by 25% in the radial direction. The stent was degraded by about 25% after 180 days. Although the parameters of the stent are relatively good, in fact, the mechanical properties of the stent cannot match the BRSs under clinical tests. In addition, the materials used in the stents are not FDA-approved. Although it is claimed that the modified material has good biocompatibility, there is still much work to be done before clinic test.

In 2017, Ware et al.[39] of Northwestern University, U.S. used bioresorbable photopolymerizable biomaterials as raw materials to print customizable BRSs with micro CLIP. This stent meet the micron-scale precision biomedical devices requirement. Compared to standard stereo lithography apparatus (SLA) technology, micro CLIP can reduce manufacturing time from hours to minutes. At the same time, stents produced by the micro CLIP has better surface finish and more uniform mechanical properties. The mechanical properties of micro CLIP stents can match Nitinol stents. After optimization, a 2 cm high vascular stent containing 4000 layers can be fabricated in 26.5 minutes.

Also in 2017, Cabrera et al.[40] of Eindhoven University of Technology, Netherlands printed the vascular stent using fused deposition modeling (FDM) technology. The printing device is BakerBot by Replicator, and the printed material is thermoplastic copolyester elastomer (TPC). Mechanical properties and differential scanning calorimetry (DSC) characteristics of the stent were tested, while, animal experiments showed that the stent was completely fused with the vessel wall within 8-16 weeks (Figure 3.C).

In 2018, Guerra et al.[41] of University of Girona, Spain printed the PCL vascular stent (Figure 3.D) using the FDM technology. The relationship between the thickness of the stent and the temperature of the nozzle as well as the printing speed and the nozzle moving speed was summarized by a large number of experiments. Although the stent in this article did not achieve high surface resolution, it explored the 3D printing process in vascular stents fabrication.

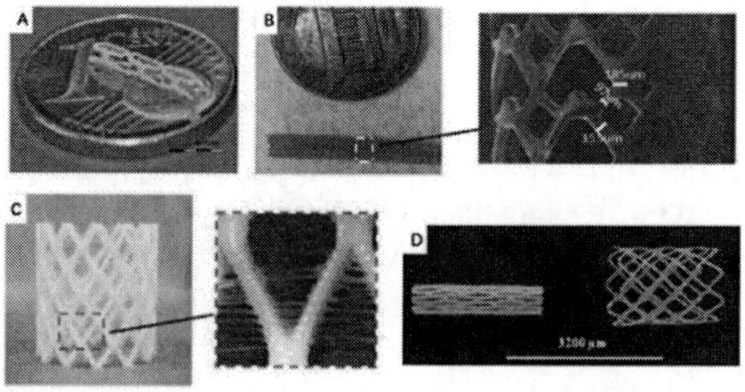

Figure 3. BRSs made by 3D printing technology. (A) BRS fabricated by Flege et al.; (B) BRS fabricated by van Lith et al.; (C) BRS fabricated by Cabrera et al.; (D) BRS fabricated by Guerra et al.

In China, Liu Yuanyuan et al.[42] of Shanghai University used a combination of bio-3D printing and electrospinning technology to form a poly(p-dioxanone) (PPDO) slide buckle stent. The team used PPDO material to prepare the inner layer of the stent by 3D printing and prepared a mixture of chitosan and poly(vinyl alcohol) (PVA) to form the outer layer of the stent by electrospinning, and planted cells on the stent. The results showed that the adhesion and proliferation of cells on stents were good because the outer layer of the stents contained chitosan, a natural biological material. This article provides a new idea for 3D-printed drug eluting and cell loading stents fabrication. Tangdu Hospital[43], Xian, successfully used the additive manufacturing technology to prepare the personalized titanium alloy porous venous extravascular stent, and completed the world's first additive manufacture extravascular stent implantation for nutcracker syndrome. Liu Qing et al.[44] developed a method for the preparation of a three-dimensional porous tubular scaffolds and equipment using additive manufacture. Lai Xuecong et al.[45] developed the additive manufacturing equipment for the preparation of cardiovascular scaffolds, which can be printed and shaped on the rotating axis, providing a new method for the preparation of vascular scaffolds.

In 2018, Wei Sun et al.[46] of Tsinghua University, China developed a 4-axis linkage 3D printing platform based on the FDM technology. In the platform an extra axis, rotating mandrel structure, is added to avoid the support structure, and developed a set of mini-screw extrusion nozzle. The nozzle can use granular biological material, and the extruded wire diameter is 100μm, which matches the requirements of the endovascular stent. The control model of the start-stop process of the screw nozzle was established to solve the start-stop delay control problem. At the same time, the effects of quenching and high-speed centrifugation on the mechanical properties of the stent were studied using PCL and PLLA materials, and the mechanical properties of the stent were optimized (Figure 4).

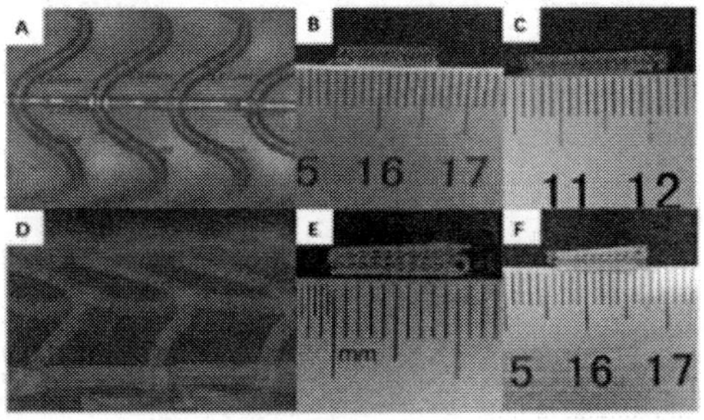

Figure 4. BRSs fabricated by Wei Sun et al., Tsinghua University. (A)Micrograph of PLLA stent; (B)Cylinder PLLA stent; (C)Conical PLLA stent; (D)Micrograph of PCL stent; (E)Cylindrical PCL stent; (F)Conical PCL stent.

3 Stent graft

3.1 Structural features of the stent graft

The stent graft is mainly used in endovascular aneurysm repair (EVAR), which has been widely used as an effective means for the treatment of descending aortic disease, aneurysm and other diseases[47]. EVAR is using a stent graft to replace the vascular wall to prevent blood entering the vascular false lumen[48]. In this way, the false lumen can be prevented from further expanding and rupturing to achieve the purpose of treatment.

The stent graft consists of a membrane and a metal stent. The film is generally made of polymer materials such as polyethylene terephthalate (PET), polytetrafluoroethylene (e-PTFE), and polyester[49]. There are also precedents for using silk as a film in China. The metal stents material are partially stainless steel, cobalt-chromium alloy and nickel-titanium alloy according to the different surgical release methods. At present, stent grafts that have been used in clinical practice are mainly produced and sold by several companies such as Medtronic, Cook, Jotec and Relay. The products use a wave or Z structure to support the blood vessels. The metal stent material is Nitinol so that self-expansion can be achieved. The film material is mainly PET and e-PTFE (Table 1).

Table 1. Stent Grafts on sale

Brand	Structure	Film Material	Metal Material
Medtronic	wave	PET	Nitinol
Jotec	wave	e-PTFE	Nitinol
Relay	wave	e-PTFE	Nitinol
Boston Scientific	wave	PTFE	Nitinol
Cook	Z	e-PTFE	stainless steel

3.2 Additive manufacturing of stent graft

At present, the mainstream stent grafts on the market include two parts: a polymer film and a nitinol stent. The forming process of Nitinol includes laser cutting, weaving, etc., which is similar to BMSs forming process[50]. While the film forming process is mainly divided into: textile forming and others. The textile forming method is to use the loom to weave polyester fiber into a tubular film. Because of its compact structure, stable size and controllable thickness, it is widely used by manufacturers of stent graft[51]. Other forming methods mainly include extrusion forming, electrostatic spinning, micro-nano composition, etc., to form PET or e-PTFE materials[52]. Medal stent and film binding methods include sewing, weaving, etc.[53]

Due to the difficulty in integrating and directly forming the film and the metal stent material, Lei Zhang et al.[54] of Tsinghua University proposed rapid prototyping sacrificial core-coating forming (RPSC-CF) technique. 3D printing is used to form a water-soluble, non-biotoxic core to ensure the inner cavity structure of the stent. Next, one or more material deposition processes, such as infiltration coating, spraying, electrostatic copy wire, are adopted to form multi-layer pipe wall structure by layer depositing materials. Finally, the inner core was immersed to obtain the stent. RPSC-CF has successfully prepared curved and bifurcation stent grafts to meet the real anatomical structure of the thoracic descending aorta. Meanwhile, the thickness of the film is controllable, which provides a basis for the patient's personalized use. The preparation cycle is about 3-4 days, and with the continuous adjustment of the preparation process, the entire stent graft fabrication is expected to be compressed within 48 hours, which can meet the treatment timing of patients with acute aortic dissection (Figure 5).

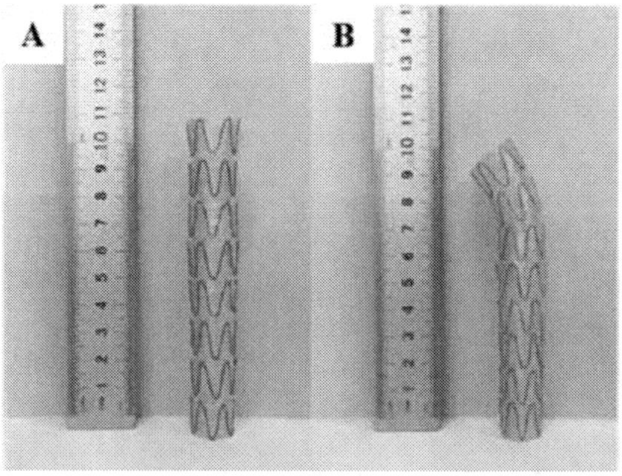

Figure 5. Stent grafts fabricated by Lei Zhang et al., Tsinghua University. (A)Straight type stent graft; (B) Curved stent graft.

4 Conclusion

Although the current vascular stent additive manufacturing technology is still in the research and development stage, the additive manufacturing technology can complete personalized stents in a short time, at the same time, avoid the treatment failure due to poor adherence and compliance. Compared with other personalized stent manufacturing processes, it has a higher application prospect in the field of clinical treatment.

With the development of precision medicine, personalized vascular stents that conform to the patient's true vascular structure have become the purpose of the next generation stents. Additive manufacturing of vascular stents has broad application prospects in the field of preparation of implantable medical devices due to its advantages of individualization and rapid prototyping. Additive manufacturing vascular stents can custom-make parameters of the stent shape, material, thickness, radial strength, etc. In 2018, Tsinghua University first packaged the biosensor on the 3D printing personalized stent graft. After the implantation, it could monitor the intravascular physiological signal in real time, track the working state of the stent stably, and provide precision treatment through data analysis. The program realizes the intelligence of the vascular stent production and provides a new idea for the development of the implantable medical device. It is believed that in the near future, a new generation of vascular stents combined with individualization, digitization, intelligence and networking will appear and provide precise medical treatment for patients with cardiovascular and cerebrovascular diseases.

Acknowledgements

Authors would like to thank Beijing Municipal Health Planning Commission (2016-2-4121), Beijing Life Science Frontier Technology Cultivation Project (Z141100000214012) and Natural Science Foundation of Beijing Municipality (Z150001).

Reference

1. Fuster, V., Top 10 cardiovascular therapies and interventions for the next decade. Nature Reviews Cardiology, 2014. 11(11): p. 671-683.
2. Li, M., A novel technique of endovascular treatment for the cerebral aneurysm: the development of covered stent specially designed for intracranial vasculature. Journal of Interventional Radiology, 2010. 19(4): p. 253-256.
3. Guo, Y. and Z. Tan, Features of endovascular stent materials and their therapeutic effects on ischemic stroke. Journal of Clinical Rehabilitative Tissue Engineering Research, 2013. 17(12): p. 2257-2264.
4. Mei, M., Characteristics and clinical application of different types of cerebral vascular scaffolds. Journal of Clinical Rehabilitative Tissue Engineering Research, 2014.
5. Liu, D., Q. Luo, and Z. Chang, Current Status and Progress of Minimally Invasive Medical Devices. Chinese Journal of Medical Instrumentation, 2008. 14(3): p. 2-11.
6. Liu, D. and T. Xi, Current Situation and Development Tendency of Interventional Medical Device&Material Industry. Chinese Journal of Medical Instrumentation, 2006. 12(12): p. 1-14.
7. White, G.H., et al., Endoleak as a complication of endoluminal grafting of abdominal aortic aneurysms: classification, incidence, diagnosis, and management. Journal of Endovascular Surgery the Official Journal of the International Society for Endovascular Surgery, 1997. 4(2): p. 152.
8. Coppi, G., et al., Experience with the Stentor endograft at four Italian centers. J Endovasc Surg, 1998. 5(3): p. 206-215.
9. Roos, J.E., et al., Detection of endograft fractures with multidetector row computed tomography. Journal of Vascular Surgery, 2005. 42(5): p. 1002-1006.
10. Junnarkar, S.P., et al., Endovascular repair for acute symptomatic and ruptured abdominal aortic aneurysms. Ulster Medical Journal, 2001. 70(1): p. 51-53.
11. Duda, S.H., et al., Physical Properties of Endovascular Stents: An Experimental Comparison. Journal of Vascular & Interventional Radiology, 2000. 11(5): p. 645-654.
12. Sigwart, U., et al., Intravascular stents to prevent occlusion and restenosis after transluminal angioplasty. N Engl J Med, 1987. 316(12): p. 701-706.

13. M, F., et al., A new austenitic stainless steel with negligible nickel content: an in vitro and in vivo comparative investigation. Biomaterials, 2003. 24(27): p. 4929-4939.
14. Jr, H.D., In-stent restenosis. Reviews in Cardiovascular Medicine, 2001. 2(3): p. 115-119.
15. Pfisterer, M., et al., Late clinical events after clopidogrel discontinuation may limit the benefit of drug-eluting stents: an observational study of drug-eluting versus bare-metal stents. Journal of the American College of Cardiology, 2006. 48(12): p. 2584-2591.
16. Johnson, J. CYPHER™ Sirolimus-eluting Coronary Stent on RAPTORTM Over-the-Wire Delivery System. 2003; Available from: https://www.accessdata.fda.gov/cdrh_docs/pdf2/p020026c.pdf.
17. Scientific, B. TAXUS Liberte and TAXUS Liberte Atom. 2008; Available from: https://www.accessdata.fda.gov/cdrh_docs/pdf6/p060008s008c.pdf.
18. Mauri, L., et al., Stent thrombosis in randomized clinical trials of drug-eluting stents. N Engl J Med, 2007. 356(10): p. 1020-1029.
19. Abbott. The XIENCEV Everolimus Eluting Coronary Stent System Instructions for Use. 2008; Available from: https://www.accessdata.fda.gov/cdrh_docs/pdf7/p070015c.pdf.
20. Rasmussen, K., et al., Efficacy and safety of zotarolimus-eluting and sirolimus-eluting coronary stents in routine clinical care (SORT OUT III): a randomised controlled superiority trial. Lancet, 2010. 375(9720): p. 1090-1099.
21. Dave, B., Bioresorbable Scaffolds: Current Evidences in the Treatment of Coronary Artery Disease. Journal of Clinical & Diagnostic Research Jcdr, 2016. 10(10): p. OE01.
22. Caiazzo, G., et al., Absorb bioresorbable vascular scaffold: What have we learned after 5 years of clinical experience? International Journal of Cardiology, 2015. 201: p. 129-136.
23. Shrivastava, S., Medical device materials. Advanced Materials & Processes, 2004.
24. Moliterno, D.J., Healing Achilles--sirolimus versus paclitaxel. New England Journal of Medicine, 2005. 353(7): p. 724-7.
25. Huang, M.H., S. Li, and M. Vert, Synthesis and degradation of PLA–PCL–PLA triblock copolymer prepared by successive polymerization of ε-caprolactone and -lactide. Polymer, 2004. 45(26): p. 8675-8681.
26. Jiang, w., et al., Application of L-lactic acid and its derivatives in food industry. Journal of Food
27. Tsuji, H., Poly(l-Lactide). 2015: Springer Berlin Heidelberg.
28. Bährlerapp, M., Springer Lexikon Kosmetik und Körperpflege. 2007, Springer Berlin Heidelberg.
29. Abbott. Absorb GT1™ Bioresorbable Vascular Scaffold (BVS) System 2016; Available from: https://www.accessdata.fda.gov/cdrh_docs/pdf15/p150023b.pdf.
30. Dun, S., Study on Micro Injection Molding Technology of Biodegradable Stent. 2013, Dalian University of Technology.
31. Huang, B., et al., Fabricating polymer stents with injection molding. 2008, US.
32. Yu, J.T., et al., Additive Manufacturing of Patient-Customizable Scaffolds for Tubular Tissues Using the Melt-Drawing Method. Materials, 2016. 9(11): p. 893.
33. T, Y., et al., Bifurcation lesions: two stents versus one stent--immediate and follow-up results. Journal of the American College of Cardiology, 2000. 35(5): p. 1145-1151.
34. Peng, H., The Numerical Analysis on the Influence of Hemodynamic to Different Cross-sections of Struts. Journal of Biomedical Engineering Research, 2013(3): p. 158-161.
35. Flege, C., et al., Development and characterization of a coronary polylactic acid stent prototype generated by selective laser melting. Journal of Materials Science Materials in Medicine, 2013. 24(1): p. 241-255.
36. Su, A.P., et al., In vivo evaluation and characterization of a bio-absorbable drug-coated stent fabricated using a 3D-printing system. Materials Letters, 2015. 141: p. 355-358.
37. Tumbleston, J.R., et al., Additive manufacturing. Continuous liquid interface production of 3D objects. Science, 2015. 347(6228): p. 1349-1352.
38. Van Lith, R., et al., 3D-Printing Strong High-Resolution Antioxidant Bioresorbable Vascular Stents. Advanced Materials Technologies, 2017. 1(9): p. 1600138.
39. Ware, H.O.T. and C. Sun. Process development for high-resolution 3D-printing of bioresorbable vascular stents. in SPIE OPTO. 2017.

40. Cabrera, M.M., et al., Computationally designed 3D printed self-expandable polymer stents with biodegradation capacity for minimally invasive heart valve implantation:A proof-of-concept study. 2017.
41. Guerra, A.J. and J. Ciurana, 3D-printed bioabsordable polycaprolactone stent: The effect of process parameters on its physical features. Materials & Design, 2017. 137.
42. Liu, Y., et al., Composite bioabsorbable vascular stents via 3D bio-printing and electrospinning for treating stenotic vessels. Journal of Southeast University, 2015(2): p. 254-258.
43. Tangdu Hospital completed the world's first 3D printed extravascular stent implantation for Nutcracker Syndrome. Journal of The Fourth Military Medical University, 2016(1): p. 4-4.
44. Liu, Q., Z. Wang, and Y. Liu, Method for the preparation of a three-dimensional porous tubular scaffolds and equipment. 2015: CN.
45. Xu, M., et al., Fast light curing nozzle for 3D printing equipment. 2015: CN.
46. Kang, S., Research on 3D printing equipment and forming technology of personalized biorebsorblable stent. 2018, Tsinghua University.
47. Anagnostopoulos, C.E., M.J.S. Prabhakar, and C.F. Kittle, Aortic dissections and dissecting aneurysms. American Journal of Cardiology, 1972. 30(3): p. 263-73.
48. Wang, S., et al., Short-term curative effect of endovascular stent-graft treatment for aortic diseases in China: a systematic review. Plos One, 2013. 8(8): p. e71012.
49. Jamshidi, P., K. Mahmoody, and P. Erne, Covered stents: a review. International Journal of Cardiology, 2008. 130(3): p. 310-318.
50. Chen, X., Research of Customized Aortic Stent Graft Manufacture. 2017, Tsinghua University.
51. Li, R., Structure analysis of different textile-based artificial vascular. Progress in Textile Science & Technology, 2007(6): p. 17-17.
52. Wang, F., Desigh and Properties of Branched Graft for Endovascular Prostheses Application. 2014, Donghua University.
53. Wang, L., Design and Process Investigation on a New Type of Thoracic Stent. 2013, Dalian University of Technology.
54. Zhang, L., Fabrication and fundamental research of multi-branched blood vessel scaffolds with multi-layered wall. 2006, Tsinghua University.

The Influence of Long-term Ageing on the Microstructure of Sanicro 25 Steel

Marek Sroka[1,a], Adam Zieliński[2], Grzegorz Golański[3] and Marek Kremzer[1]

[1] *Silesian University of Technology, Institute of Engineering Materials and Biomaterials, 18A Konarskiego Str., 44100 Gliwice, Poland*
[2] *Institute for Ferrous Metallurgy, K. Miarki 12-14, 44100 Gliwice, Poland*
[3] *Czestochowa University of Technology, Institute of Materials Engineering, Al. Armii Krajowej 19, 42200 Czestochowa, Poland*

Abstract. Sanicro 25 (22Cr25NiWCoCu) is a newly developed austenitic stainless steel used in the boiler pressure superheaters and reheaters with ultra-supercritical high-efficient parameters. The paper presents the results of microstructure testing and secondary phase separation processes after 20,000 h ageing at 700 and 750 °C. In the initial stage of ageing, the precipitation of numerous very small $M_{23}C_6$ chromium carbides and Laves phase were observed. The long-term ageing related to temperature revealed the existence of secondary phases: $M_{23}C_6$, MX, NbCrN, Laves phase and σ phase. The test results presented in the paper are part of the developed material characteristics to be used for predicting the service life of the material of components operated under creep conditions.

1 Introduction

In spite of the extensive promotion of renewable energy sources (solar, wind or water) for years and the expected increase in the share of nuclear technologies in the energy balance, not only in Europe but also in the world power industry, the global energy sector is still mostly based on fossil fuels.

Despite the significant rise in prices of harmful gas emissions into the air, which is to curb the development of the conventional power generation, there are no signs that the proportion of sources of electricity generation could significantly change in the coming years [1-3]. The electricity generation in Poland is still based on coal and lignite. Due to their considerable reserves, they are the basic and at the same time the cheapest source of energy [1-3]. Therefore, to reduce the costs of electricity generation, the obsolete power units (mostly with a capacity of 200 MW, built a few decades ago) have been modernised and the modern supercritical power units have been built for several years in Poland. It is inextricably linked to the development of the materials technology and the need to use materials with increased performance. Other important aspects are the optimisation of processes and installation modules, strict control of the operation, the increase in thermal flexibility and the introduction of the carbon capture technology [4-7].

[a] Corresponding author : marek.sroka@polsl.pl

One of the most modern currently used materials is Sanicro 25 (X7NiCrWCuCoNbNB25-23-3-3-3-2) steel, which was developed by AB Sandvick Material Technology. This steel is characterised by high nickel (24%), chromium (22%) and tungsten (3%) contents and the addition of elements such as niobium, cobalt and copper. It is primarily intended for components of boiler pressure superheaters and reheaters to be operated at up to 700 °C [8].

The Sanicro 25 has higher creep strength compared to other austenitic structural steels, i.e. HR3C and Super 304H. Creep strength of Sanicro 25 steel determined at 700 °C for up to 100,000 h is 95 MPa, and it is similar to that of the much more expensive HR6W nickel-base alloy [1, 8, 9]. The high performance of Sanicro 25 steel justifies the increasingly higher demand for it and its more and more widespread use in ultra- supercritical and advanced ultra- supercritical power units [1, 9].

2 Material for investigations

The material used in the investigations was Sanicro 25 steel delivered in the form of a coil sample of ϕ38x8.8mm. The chemical composition of the investigated steam superheater sample is presented in Table 1. The microstructure of the Sanicro 25 steel was observed with Inspect F scanning electron microscope (SEM) on conventionally prepared electrolytically etched metallographic microsections. The metallographic examinations were carried out on the material in the as-received condition and after long-term ageing at 700 and 750 °C for 10,000 and 20,000 h.

Table 1. Chemical composition of investigated material, wt.%

Chemical composition, wt.%												
C	**Si**	**Mn**	**P**	**S**	**Cu**	**Cr**	**Ni**	**W**	**Co**	**Nb**	**B**	**N**
0.06	0.25	0.50	0.01	<0.01	2.9	23.0	24.1	3.2	1.4	0.4	0.005	0.17

3 Results and discussions

The microstructure of Sanicro 25 steel in the as-received condition, observed with the scanning electron microscope, is shown in Fig. 1. The investigated steel is characterised by the austenitic matrix with the grain size of 7 according to the scale of models with annealing twins and numerous primary precipitates of varying size arranged in bands both within the grains and at the grain boundary. The structure of Sanicro 25 steel is typical of the niobium-stabilised heat-resistant austenitic steels.

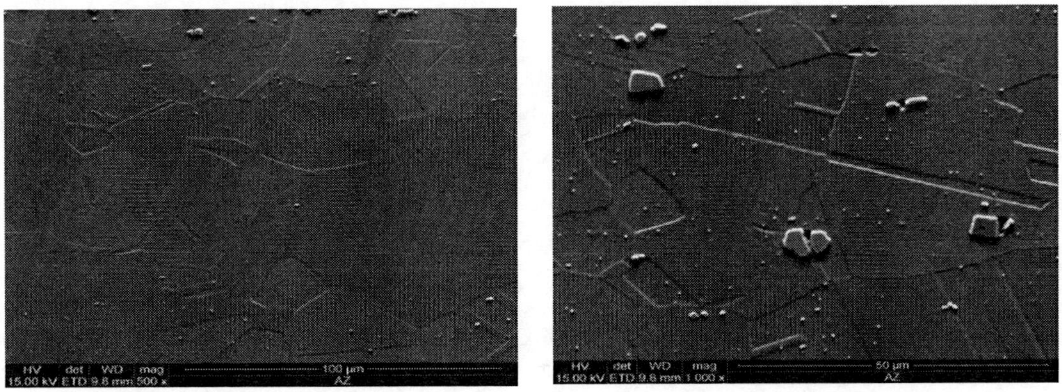

Figure 1. Microstructure of Sanicro 25 steel in the as-received condition, hardness 184 HV10

Long-term ageing of Sanicro 25 steel contributes to the secondary phase precipitation strengthening (Figs. 2, 3, 4). In the initial stage of ageing, there is an intensive precipitation process during which numerous excellent copper particles coherent with the matrix are precipitated within the grains. The copper-rich precipitates (called ε_Cu precipitates) in austenitic steels are observed just after one hour of ageing at 650 °C [10]. Due to their nanometric dimensions, they strengthen the alloy very intensively [10, 11]. The second precipitate that occurs in the structure of the investigated steel is $M_{23}C_6$ carbides. The $M_{23}C_6$ carbides precipitate mainly at the grain boundaries, and in contrast to the ε_Cu precipitates, they are characterised by lower thermodynamic stability, which results in their high tendency to coagulate and create the so-called continuous network of precipitates at the grain boundaries [12, 13].

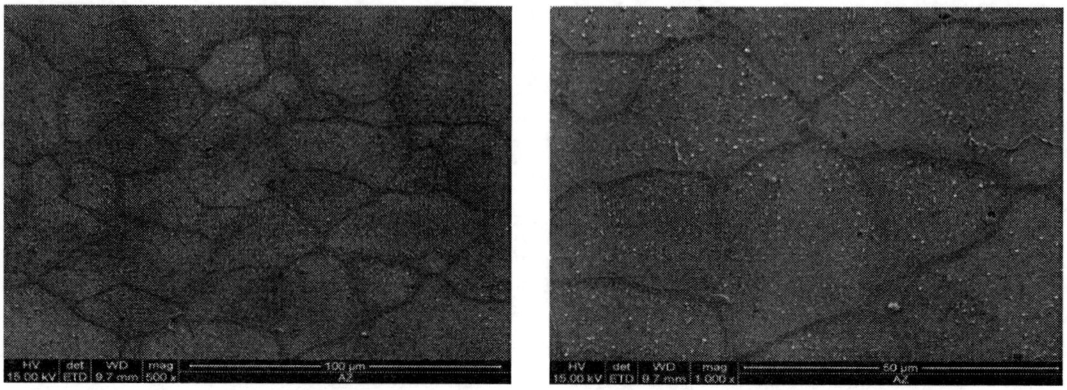

Figure 2. Microstructure of Sanicro 25 steel after 1000 h ageing at 700°C, hardness 215HV10

Figure 3. Microstructure of Sanicro 25 steel after 1000 h ageing at 750 °C, hardness 226HV10

After 1000 h ageing at both 700 and 750 °C, the fine Laves-phase precipitates (Fe_2W) were revealed in the matrix of the investigated steel (Fig. 2). The Laves-phase precipitates were observed both within the grains and at the grain boundaries. The fine particles of $M_{23}C_6$ carbides and Laves phase precipitated at the grain boundaries have a positive effect on the increase in creep strength by inhibiting the slip at the grain boundaries [12, 13]. After 10,000 h ageing, the increase in the size of precipitates of this phase was observed (Fig. 4,5). Longer ageing times of approx. 20,000 h contribute only to the slight growth of the Laves phase particles (Figs. 6, 7). The impact of the Laves phase precipitated within the grains on steel properties depends on its size and volume fraction. The finely dispersed Laves phase particles precipitated within the grains have a positive effect on creep and fatigue strength. The Laves phase coagulation during service results in the gradual disappearance of this positive effect [9, 14].

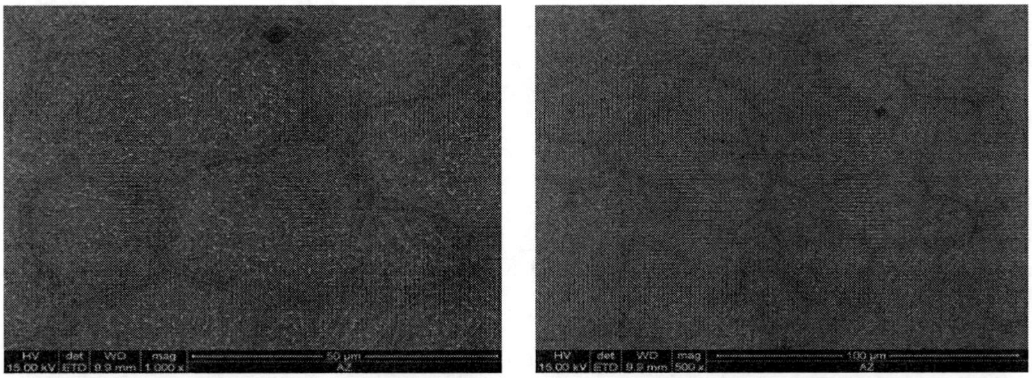

Figure 4. Microstructure of Sanicro 25 steel after 10,000 h ageing at 700 °C, hardness 234 HV10

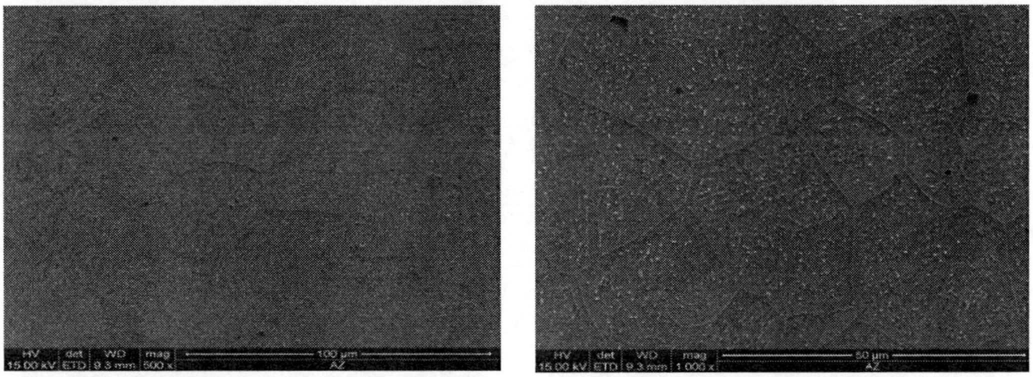

Figure 5. Microstructure of Sanicro 25 steel after 10,000 h ageing at 750 °C, hardness 247 HV10

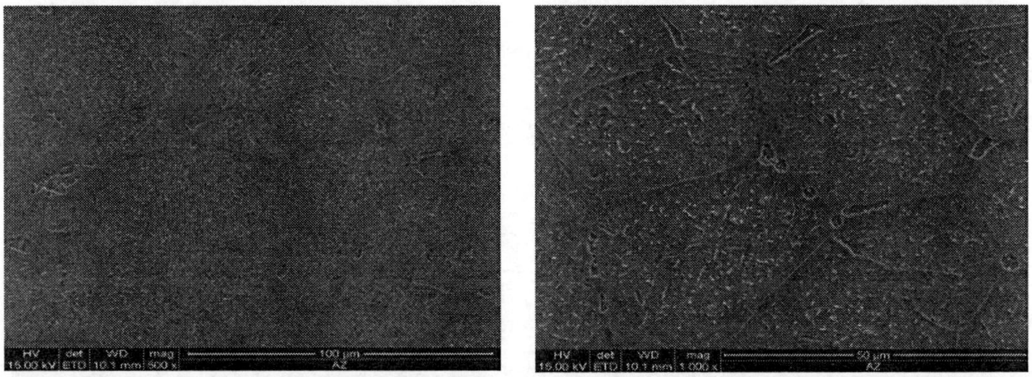

Figure 6. Microstructure of Sanicro 25 steel after 20,000 h ageing at 700 °C, hardness 249 HV10

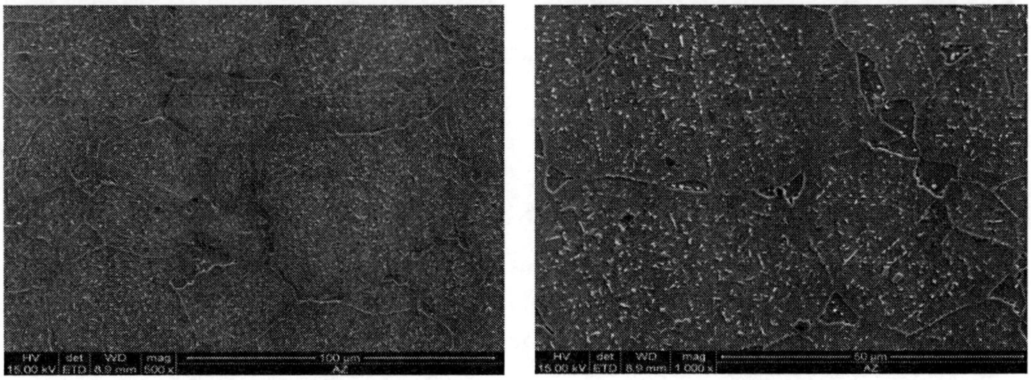

Figure 7. Microstructure of Sanicro 25 steel after 20,000 h ageing at 750 °C, hardness 239 HV10

Long-term ageing also contributed to the precipitation of unfavourable σ phase at the austenite grain boundaries. The privileged place was the meeting point of the three-grain boundaries (Figs. 6, 7). The precipitation of the σ phase results in a significant reduction in plastic properties and ductility of austenitic steels [9, 12, 13].

4 Summary

The metallographic examinations carried out for Sanicro 25 steel in the as-received condition and after ageing at 700 and 750 °C have shown that:

- in the as-received condition, Sanicro 25 steel is characterised by the austenitic microstructure with numerous primary MX and Z-phase (NbCrN) precipitates;
- ageing leads to the precipitation of copper-rich (ε_Cu) particles, Laves phase and secondary Z-phase precipitates in the matrix, while at the grain boundaries, the occurrence of $M_{23}C_6$ carbides and the Laves- and σσ-phase precipitates were observed.
- the extension of the ageing time up to 20,000 h at the test temperature results in an increase in the number and size of precipitates both within the grains and at the grain boundaries, and it is particularly visible in relation to the $M_{23}C_6$ carbides and σ phase.

Acknowledgements

This research was financed by National Science Centre, 2011/01/D/ST8/07219 – Project: "Creep test application to model lifetime of materials for modern power generation industry" and co-financed within the framework of the statutory financial grant supported by the Faculty of Mechanical Engineering of The Silesian University of Technology.

References

1. A. Hernas, A. Dobrzański, J. Pasternak, S. Fudali, *Characteristics of a new generation of materials for the power industry* (Publishing House of Silesian University of Technology, Gliwice, 2015)
2. M. Sroka, A. Zieliński, J. Mikuła, Arch. Metall. Mater. **61**, 1315 (2016)
3. T. Tański, M. Sroka, A. Zielinski, *Creep* (InTech, Rijeka, Croatia, 2018)
4. A. Zieliński, M. Sroka, A. Hernas, M. Kremzer, Arch. Metall. Mater. **61** 761 (2016)
5. A. Zieliński, J. Dobrzanski, H. Purzynska, G. Golanski, *Mater. Test.* **57**, 859 (2015)
6. M. Sroka, A. Zieliński, A. Hernas, Z. Kania, R. Rozmus, T. Tanski, A. Śliwa, Metalurgija **56**, 333 (2017)
7. G. Golański, G., A. Zielińska-Lipiec, A. Zieliński, M. Sroka, J. Mater. Eng. Perform. **26**, 1101 (2017)
8. A. Zieliński, J. Dobrzański, H. Purzyńska, R. Sikora, M. Dziuba-Kałuża, Z. Kania, Arch. Metall. Mater. **62**, 2057 (2017)
9. G. Golański, *High-temperature austenitic steels* (Publishing House of the Faculty of Production Engineering and Materials Technology of the Czestochowa University of Technology, Częstochowa, 2017)
10. J.W. Bai, P.P. Liu, Y.M. Zhu, C.Y. Chi, X.Y. Yu, X.S. Xie, Q. Zhan, *Mat. Sci. Eng. A-Struct.* **584**, 57 (2013)
11. Ch.–Y. Chi, H.–Y. Yu, J.–X. Dong, X.–S. Xie, X.–F. Chen, F. Lin, Acta Metall. Sin. **24**, 141 (2011)
12. T. Sourmail, Mater. Sc. Techn. **14**, 1 (2001)
13. J.-Z. Wang, Z.-D. Liu, H.-S. Bao, S.-C. Cheng, B. Wang, J. Iron Steel Res. Int. **20**, 54 (2013)
14. G. Chai, M. Bostrom, M. Olaison, U. Forsberg, Procedia Mater. Sc. **55**, 232 (2013)

The Corrosion Studies of Electron Beam Welded Nickel Alloy

M. Sroka[1], E. Jonda [1,a], M. Węglowski[2] and S. Błacha[2]

[1] *Silesian University of Technology, Institute of Engineering Materials and Biomaterials, 18A Konarskiego Str., 44100 Gliwice, Poland*
[2] *Institute of Welding, 16/18 Bł. Czesława Str., 44100 Gliwice, Poland*

Abstract. The paper presents the influence of electron - beam (EB) remelting effect on the surface layer electrochemical parameters obtained from potentiodynamic anodic polarization studies and impedance spectroscopy measurements for a set of Inconel 617 electron beam remelted obtained for different process parameters. The correlation between EBW process parameters and characteristic of surface oxide layer properties and resistance to the acidic environment were discussed. The electrochemical studies were supported by microstructural analysis of the remelted zone (RZ), heat affected zone (HAZ), native metal and observed precipitates formed under rapid solidification process. Both electrochemical technics applied to evaluate corrosion properties of remelted Inconel 617 evidenced a strong influence of the electron beam current on the corrosion resistance.

1 Introduction

The studies conducted up to now have been mainly focused on the technology of joining nickel alloys by EBW welding. The EB remelting process is still weakly studied in the literature. The main problems of welding considering nickel base alloys include the tendency to hot cracking of crystallization and liquation type, wherein the crystallization cracks can develop in the weld axis or may be arranged in the transverse direction and also may be present as a micro - cracks. The liquation cracks are mainly located in the near weld area of heat affected zone (HAZ). The cause of hot cracks are the liquid phases present on the grain boundary with a lower melting temperature, along which occurs the metal decohesion due to shrinkage stresses during crystallization and cooling of the weld [1-7].

From the view point of the corrosion resistance of an EB remelted layer, severe galvanic corrosion could take place in the fusion zone or HAZ due to their very small anodic areas when the corrosion potential of both zones is more active than the adjacent base material [8]. Occurring corrosion processes are always connected to microstructural inhomogeneity resulting from alloying element partitioning between present phase and formation of complex precipitates [9-10]. Although excellent microstructural features, heat and mechanical properties of electron beam weldments, corrosion resistance is also one of the most important factors in evaluation of long term life service of Inconel

[a] Corresponding author : ewa.jonda@polsl.pl

alloys. Many authors postulated that evaluation of corrosion resistance is absolutely essential to prevent premature damages of weldments [11-13].

In the present paper, the influence of basic electron beam remelting process parameters on the resistance to the acidic environment and selected surface properties were discussed.

2 Experimental conditions

In the present study, the Inconel alloy 617 has been electron beam remelted. The electron beam (EB) remelting of Inconel 617 alloy has been performed on the universal EB welding system type XW150:30/756 (Cambridge Vacuum Engineering). The base material Inconel 617 alloy was supplied in the form of a solution treated and a water quenched plate of the thickness of 1.0 mm. Six weld beads were produced on the surface with parameters: beam current: 1 mA, 1.5 mA, 2 mA, 2.5 mA, 3mA, 3.5 mA; accelerating energy: 120 kV, remelting velocity: 500 mm/min., and focus: 496 mm. The corrosion behaviors of welded nickel 617 alloy were evaluated by analyzing the polarization curves and electrochemical impedance spectroscopy. Metallographic investigations were made with the Leica light microscope at magnifications 10 – 1000x. The metallographic studies were performed on the scanning electron microscope Zeiss Supra 35. Analysis of the chemical composition was performed using an ARL3460 spectrometer. Specimens of remelted Inconel 617 were observed also under TEM in order to midentify present precipitates. The TEM experiments were performed on the TITAN FEI 80 – 300KV Probe Corect.

3 Results and discussion

The electron beam remelting of the INCONEL 617 surface layer resulted in the formation of three different zones: the remelted zone (RZ), the heat affected zone (HAZ) and the base metal (BM) one. The so - called fusion line also was formed by melting of grains of HAZ (Fig. 1).

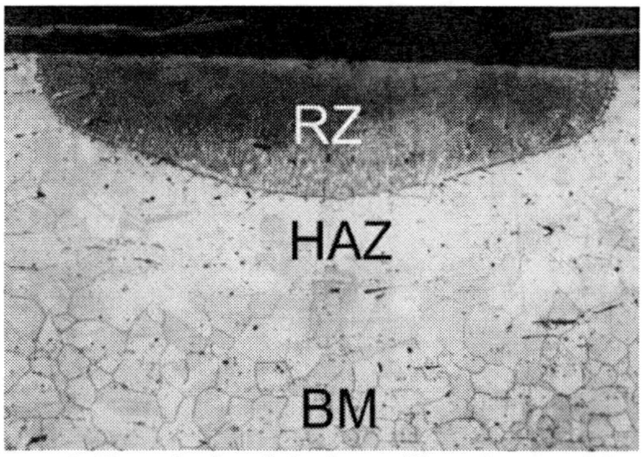

Figure 1. Remelted and the heat affected zone of the surface layer of INCONEL 617, beam current: 1 mA.

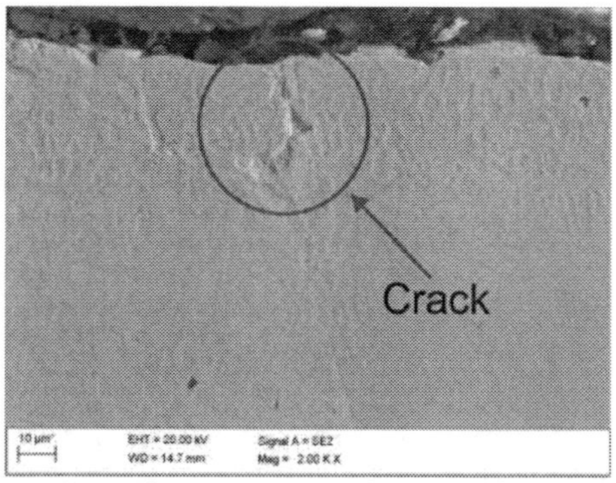

Figure 2. Microstructure of Inconel 617 alloy remelting by electron beam with a visible crack in remelting zone, beam current 1.5 mA.

The geometrical parameters of the weld bead have been studied. The lowest width of remelted weld face of 821.71 μm was obtained for EB remelting at the beam current of 1 mA, while the highest 2479.78 μm for welding current 3.5 mA. Analogically the minimum remelting depth of 193.4 μm was achieved at 1 mA beam current, while the highest 1137.27 μm when 3.5 mA beam current was applied. It can be concluded that the beam current value has a significant influence on the remelted weld face width and remelting depth and its increase results in an increase of both parameters. The microstructure of EB remelted Inconel 617 reveled presence of singular cracks that extends transversely to the axis of the remelted zone (RZ). Microstructure made by electron beam remelting with a visible crack in remelting zone is shown in Figure 2 . Due to the place of origin, it can be concluded that it is a hot crack, caused by rapid solidification process. Such crack may occur because crystallized metal is under the influence of the tensile stresses arising from the uneven cooling of the heated material. Because of present stresses, the metal deforms, and insufficient capacity to deformation causes it's cracking. Hot cracks are one of the most common defects that occur during the surface modification by electron beam or after a certain time after the treatment, but before external stresses from service loads [8].

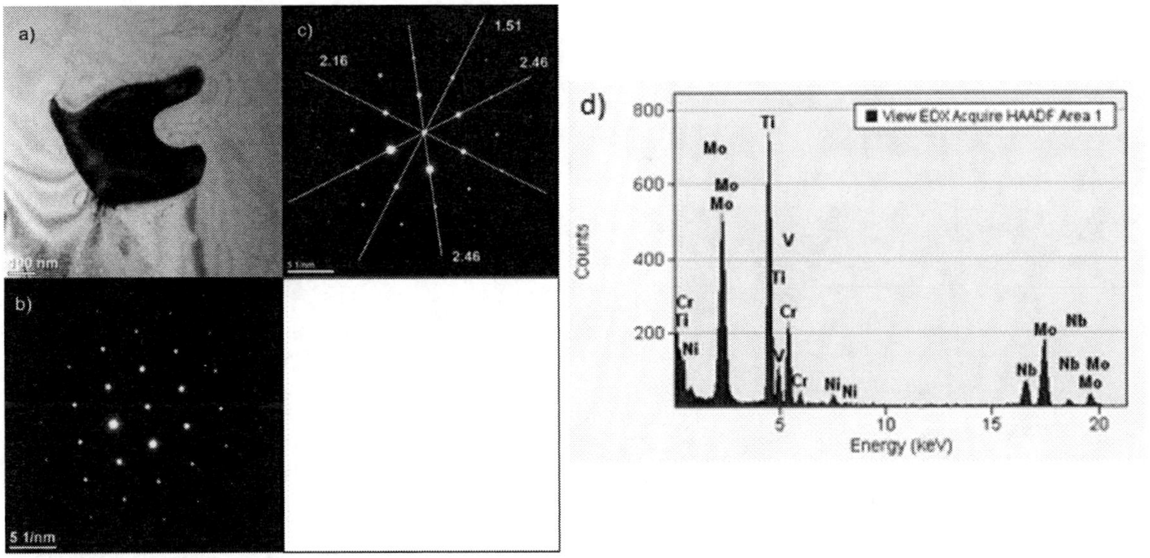

Figure 3. The thin foil structure of Inconel 617 alloy after EB remelting, beam current 3 mA; a) bright field, b) diffraction pattern of the area showed in fig a) and c) solved diffraction of the area showed in fig a), d) spectrum obtained by EDS analysis.

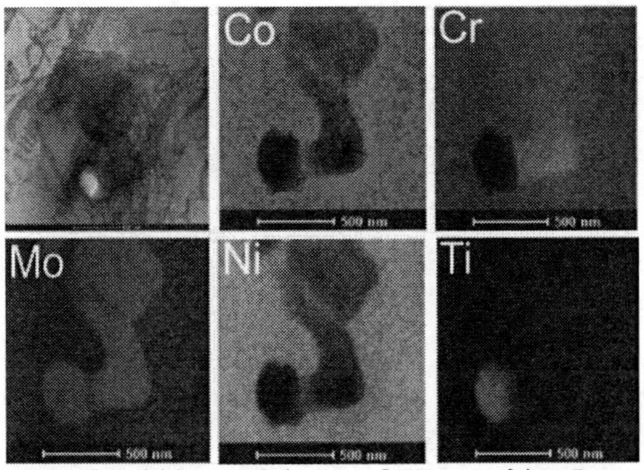

Figure 4. Elemental map distribution of carbide precipitates of EB remelting Inconel 617 alloy (at beam current of 3 mA), (HAADF – The High Angle Annular Dark Field detector).

Based on the thin foil studies in a transmission electron microscope it was found, that in the electron beam remelted material (at the beam current of 3 mA) is the precipitates of (Mo, Ti)C_2. The orientation of the precipitates with respect to the beam was [01-1] (Fig. 3). Precipitates of (MoTi)C_2 can be checked in the card number 01-071-6035. The results of mapping of the chemical elements distribution (Fig. 4) also shows the significant increase of Ti and Mo and smaller of Cr in vicinity of precipitates concentration in comparison to the base material.

The microstructural studies on SEM revealed that the morphology of EB remelted zone shows cellular - dendritic microstructure. The dendritic growth occurs according to the direction of heat dissipation from the border of the remelted zone (RZ) and the heat affected zone (HAZ). The microstructure of the remelted zone are presented in Figures 5-6. The remelted zone shows columnar crystals and dendritic cells crystallizing in the direction of heat dissipation. The microstructure of EB remelted Inconel 617 alloy at low beam power (1 mA) shows typical dendritic microstructure of fast crystalizing alloy, where formed cellular - dendritic grains are small and preferentially more rectangle

 https://doi.org/10.1051/matecconf/201925303005

(Fig. 6 a). In the case of higher beam power during EB remelting (i.e. 3.5 mA) formed grains in remelted zones are preferentially more elongated and bigger than for lower remelting energies (Fig. 6 b). Performed microstructural analysis not reveled any other precipitates in interdendritic regions, proving full dissolution of primary carbides during remelting and lack of their precipitation during cooling form remelting temperature.

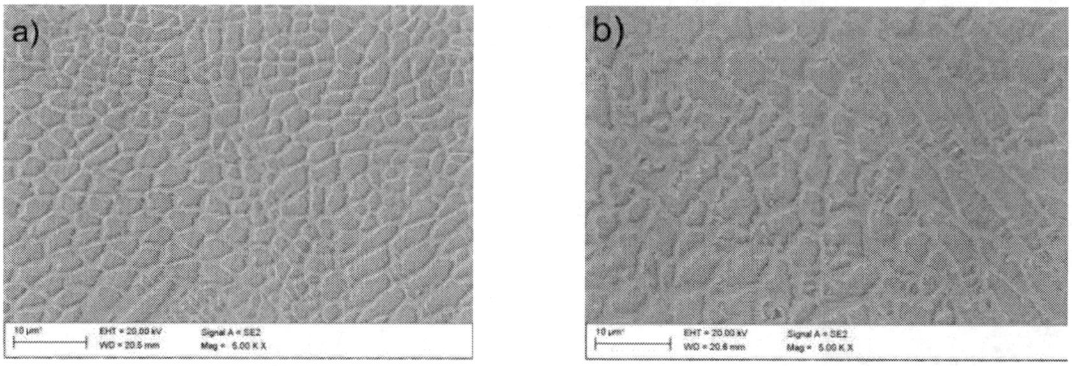

Figure 5. SEM micrographs of Inconel 617 alloy, beam current: a) 1 mA, b) 2.5 mA.

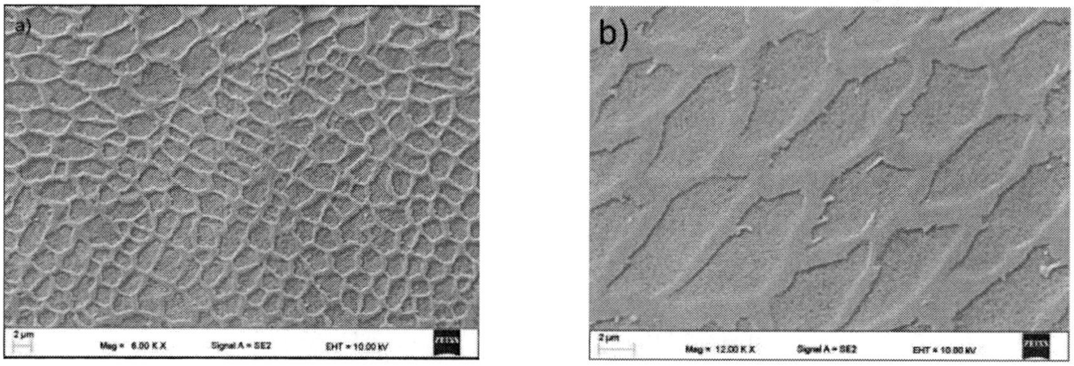

Figure 6. The microstructure in the centre of remelted zone of Inconel 617 alloy remelted with electron beam power of: a) 1 mA b) 3.5 mA.

Both electrochemical technics applied to evaluate corrosion properties of remelted Inconel 617 evidenced a strong influence of the electron beam current on the corrosion resistance. It must be remembered that during the potentiodynamic test the sample surface got alteration while the EIS method introduces smaller perturbation to tested system. The EIS parameters analysis and Nyquist and Bode plots confirmed the decreasing resistance character (R2 parameter) of the remelted surface, thus worst passivation ability in the test 1M NaCl solution. The Nyquist plots (Fig. 7a) of EIS data show changes in the slope of the impedance Z curve of the remelted surfaces. The general trend of decreasing impedance with the increase of applied beam current was observed, indicating thickening of the passive layer lower resistant character.

The best curve characteristic was registered for a parent metal, not EB remelted. The EIS Bode phase angle plots (Figs. 7b and 7c) shows high phase angles over an intermediate frequency range with sudden follow at high rang for the parent metal, typical for a passivated surface. The Bode plots show typical capacitive like behaviour, where the phase angle maximums are smaller than 80° what can be interpreted as a deviation from ideal capacitator behaviour. The Bode plot analysis confirms the resistive nature of studied surfaces, as evidenced by the phase angle value close to 0 and almost constant value of log Z in the high-frequency range. The Bode characteristic was changed after EB remelting, the phase angle curve shows high values in the high-frequency range.The potentiodynamic polarisation curves (Fig. 8) revealed a shift to positive values of corrosion potential (E_{corr}) for small electron beam current welds (beam current 1.5 - 2.5 mA), while for higher (3 mA) it was returned to the values of the parent metal and exceeded it for a maximum EB energy (3.5 mA). Similarly, with EB current increase, the polarization resistance (R_p) decreased and increased corrosion current density (i_{coor}) revealing deterioration of corrosion resistance of EB remelted Inconel alloy. The increase of EB beam current resulted in gradual decrees of the repassivation potential (E_{REP}) from positive to negative values.

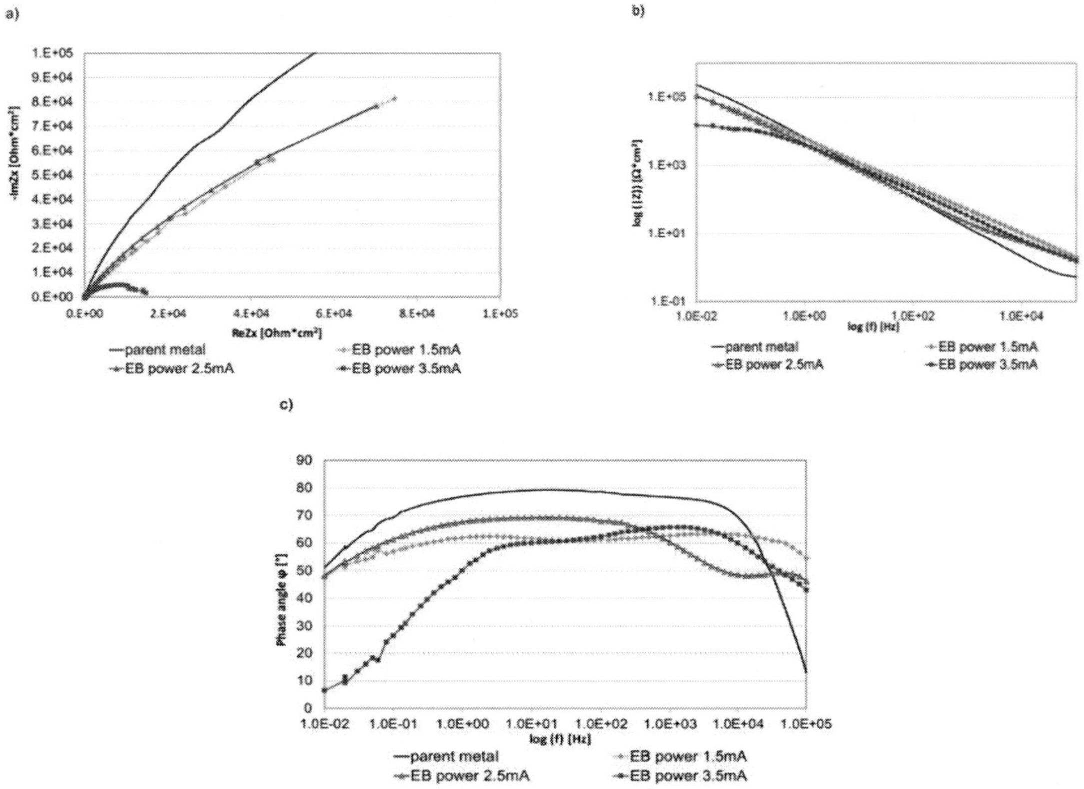

Figure 7. EIS of Inconel 617 – the parent metal and welds at different electron beam power, a) Nyquist plot, b) Z modulus and c) Bode phase-angle plot.

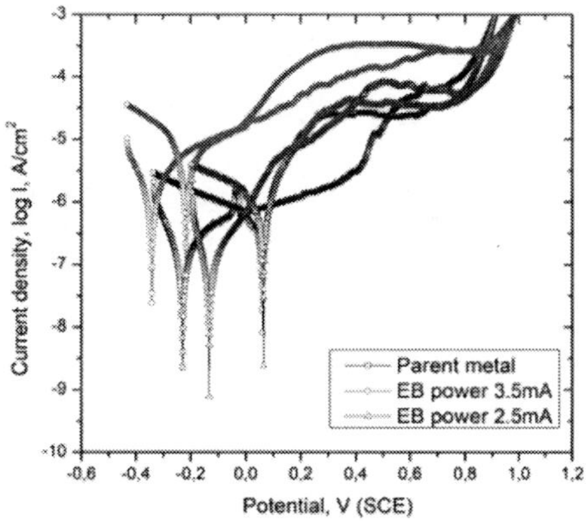

Figure 8. Potentiodynamic polarization curves of EB welded Inconel alloy 617 in 1MNaCl at different electron beam power.

4 Summary

On the basis of test results, it was found that:

1. The microstructure of remelted zone shows cellular - dendritic microstructure, where for low electron beam power formed crystals are small and rectangular that's grown with applied energy showing more elongated form.
2. During electron beam remelting of Inconel 617 alloy singular cracks were observed, that extends transversely to the axis of the remelted zone (RZ). Due to the place of origin, it can be classified as hot crack, caused by rapid solidification process.
3. With the increase of electron beam power, the thickness of the remelted zone (RZ) and width of remelted surface increased.
4. The corrosion resistance of the surface, evaluated by means of EIS and polarization curves analysis, was strongly deteriorated after EB remelting.

Acknowledgements

This research was financed within the framework of the statutory financial grant supported by the Faculty of Mechanical Engineering of The Silesian University of Technology.

References

1. J. Nowacki, K. Pietrzak, An overview of welding **11**, 22 (2012)
2. M. Lachowicz, W. Dudziński, M. Podrez - Radziszewska, Mater. Charact. **59**, 560 (2008)
3. M. Shakil, M. Ahmad, N.H. Tariq, B.A. Hasan, J.I. Akhter, E. Ahmed, M. Mehmood, M.A. Choudhry, M. Iqbal, Vac. **110**, 121 (2014)
4. J. Kusiński, J. Blicharski, M. Cieniek, Ł. Dymek, S. Rozmus-Górnikowska, M. Solecka, M. Faryj, Mater. Eng. **36**, 363 (2015)
5. T. Wang, N. Li. Z. Yhang Vac. **149**, 29 (2018)
6. K.D. Ramkumar, R. Sridhar, S. Periwal, S. Oza, V. Saxena, P. Hidad, N. Arivazhagan, Mater. Des. **68**, 158 (2015)
7. C. Wieding, C. Lochbichler, N. Enzinger, C. Beal, C. Proc. Engineering **86**, 184 (2014)
8. S. A. David, J. M. Vitek, S.S. Babu, Sci. and Techn. of Welding and Joining **2**, 79 (1997)
9. M. Lucas, JR., C.E Jackson, Welding Research Supplement, 46 (1970)
10. L.Y. Xu M. Li, H. Y. Jing, Y. D. Han, Int. Jour. Electrochem. Sci. **8**, 2069 (2013)
11. K. Łyczkowska, J. Michalska, Arch. of Met. and Mat. **6 (2)**, 653 (2017)
12. G. Fuyang, G. Peng, J. Zhiying, Z. Liu, Z.Liao, Int. Jour.of Light Weight Mat. And Man. (to be published) (2018)
13. S. Schmigalla, S. Krasnorutskyi, M. Zinke, A. Heyn, Mat. and Corrosion **66 (5)** (2015)

Oxidation Resistance of Modified Aluminide Coatings

Jolanta Romanowska[1,a], Maryana Zagula-Yavorska[1] and Łukasz Kolek[1]

[1]*Department of Materials Science, Faculty of Mechanical Engineering and Aeronautics, Rzeszów University of Technology, Al. Powstańców Warszawy 12, 35-959 Rzeszów, Poland*

Abstract. The application of protective aluminide coatings is an effective way to increase the oxidation resistance of the treated parts and prolongs their lifetime. The addition of small amount of noble metals (platinum or palladium) or reactive elements such as: hafnium, zirconium, yttrium and cerium has a beneficial effect on oxidation behavior. This beneficial effect includes an improvement of adhesion of alumina scales and reduction of oxide scale growth rate. Platinum and hafnium or zirconium modified aluminide coating were deposited on pure nickel using the electroplating and CVD methods. The coatings consisted of two layers: an outer, β-NiAl phase and the interdiffusion γ'-Ni_3Al phase. Palladium dissolved in the whole coating, whereas hafnium and zirconium formed inclusions on the border of the layers. Samples were subjected to cyclic oxidation test at 1100 °C for 200h. Oxidation resistance of the palladium, Hf+Pd and Zr+Pd modified coatings deposited on pure nickel does not differ significantly, but is better than the oxidation resistance of the non-modified one.

1. INTRODUCTION

Protective aluminide coatings, characterized by low density, high melting point and oxidation resistance, are deposited on superalloys as bond coats in thermal barrier coatings (TBC) [1]. Aluminum in the coatings reacts with oxygen from the atmosphere forming α-Al_2O_3 at the surface, which protects the substrate from oxidation and hot corrosion [2]. Oxidation of coatings degrades them through internal weakening and deformation. Slowing down the oxide scale growth as well as improvement of their adherence to the substrate will improve oxidation resistance. That has been achieved by alloying them with small amount of noble metals (platinum [3], palladium [4]) or reactive elements (hafnium [5], zirconium [6]).

Palladium addition to these coatings are effective in reducing the oxide growth rate and scale adherence [4,7]. Yavorska et al. [8] proved, that palladium modified aluminide coatings on Inconel 713 LC and CMSX 4 superalloys contain mainly the (Ni,Pt)Al, and (Ni,Pd)Al phases. Higher total content of alloying elements such as molybdenum, tungsten chromium, titanium, tantalum and rhenium in CMSX 4, as compared to Inconel 713 LC, favoured formation of thinner alumina coatings. However, hafnium is the most efficient dopant in reducing the oxide scale growth rate [6]. Hafnium segregates to the alloy-scale interface and grain boundaries. Hafnium forms nanoparticles in aluminide coatings on pure nickel [9]. Its out-diffusion along oxide grain boundaries to the oxide scale surface is slowed down by the large atomic radii. This slow diffusion stands against the aluminum

[a] Corresponding author : jroman@prz.edu.pl.

outward diffusion and oxygen inward diffusion [10,11]. This way the scale adhesion is improved and the growth rate of the scale is reduced.

Small Zr addition (about 0.1 at.%) [12] improves coatings' oxidation resistance, because of the diminishing of the scale spalling. Zirconium forms nanoparticles in aluminide coatings on pure nickel [13]. After oxidation, it is homogeneously distributed within the scale, delays spallation and improves adhesion [14].

"Co-doping" - the idea of using more than one dopant is a promising way for improvement of oxidation resistance of coatings [11,15]. Many pairs of dopants were analyzed, for instance: Y+La, Dy+Hf ,Y+Hf, Pt+Pd, , La+Hf, , Hf+Zr, or Y+La [16,17]. Oxidation resistance of co-doped coatings is better than the respective single–doped ones. However, the Pd+Hf and Pd+Zr co-doping has not been investigated yet.

Therefore, this paper presents analysis of oxidation resistance of the Pd+Hf and Pd+Zr co-doped aluminide coatings deposited on pure nickel.

2. EXPERIMENTAL

Palladium layers were deposited on nickel (of technical purity) by the electrochemical method. The precise procedure of the electrodeposition process in presented in reference [18]. Then, the aluminide coatings were deposited by the CVD method [19, 20].

Chemical and phase composition of coatings was analyzed by SEM and EDS methods. The oxidation resistance was analyzed by placing samples in the furnace, than heating them up to 1100 °C and maintaining it at this temperature for 20 hours. Than samples were cooling to the room temperature with the furnace. After each cycle samples were weighed and inspected.

3. Results and discussion

3.1 Palladium modified aluminide coating on pure nickel

Palladium modified aluminide coating on pure nickel has a double-zone structure (Fig. 1, Table 1). In the first, additive zone, on the top of the coating, the proportion of Ni to Al corresponds to the β-NiAl phase (Table 1, Point 1). Moreover, some palladium (5.2-4.5 at. %) was identified (Fig. 1-2, Table 1-2). This indicates formation of the β-(Ni,Pd)Al phase. The thickness of this zone is ~52 μm. The second, interdiffusion zone, situated below is much thinner (~5 μm). Its chemical composition suggests formation of the palladium doped γ'-Ni_3Al phase (Table 2, Point 3). The nickel substrate/palladium doped γ'-Ni_3Al as well as the palladium doped γ'-Ni_3Al/β-(Ni,Pd)Al interfaces were relatively flat (Fig. 2). Kirkendall-like porosity was observed within the β-(Ni,Pd)Al phase near palladium doped γ'-Ni_3Al phase (Fig. 2).

Table 1. Chemical composition on the cross-section of the palladium modified aluminide coating deposited on the nickel substrate (in at.%)

Spot	Chemical composition		
	Al	Ni	Pd
1	34.3	60.6	5.2
2	-	100	-

Table 2. Chemical composition on cross-section of additive/interdiffusion zone of palladium modified aluminide coating deposited on the nickel substrate (in at.%)

Spot	Chemical composition		
	Al	Ni	Pd
1	32.4	63.1	4.5
2	31.3	64.6	4.1
3	21.7	77.5	0.8
4	-	100	-

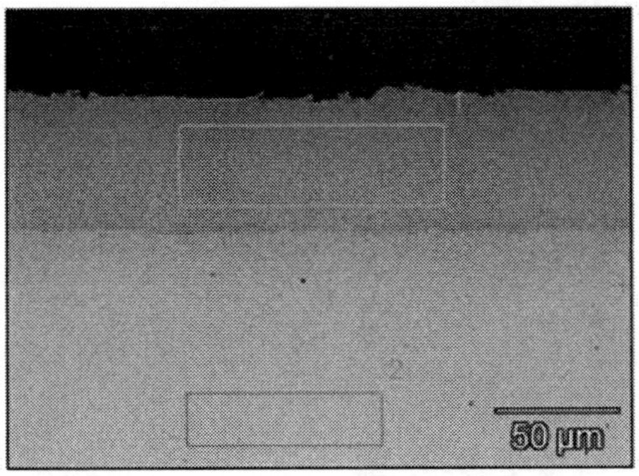

Figure 1. SEM microstructure on the cross-section of the palladium modified aluminide coating deposited on the nickel substrate.

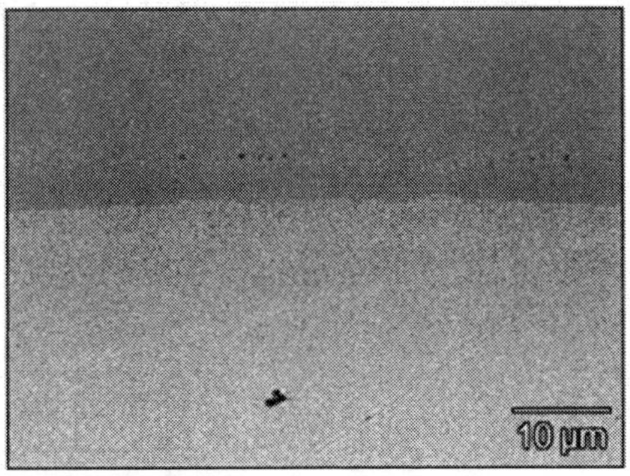

Figure 2. SEM microstructure on the cross-section of the additive/interdiffusion zone of the palladium modified aluminide coating deposited on the nickel substrate.

EDS analysis and concentration profiles of the cross-section of the palladium modified aluminide coating showed outward nickel diffusion from the substrate and the inward aluminum diffusion from the surface to the nickel substrate (Figure 3). It was found that palladium diffuses both outward from the surface to the additive zone and inward from the surface to the interdiffusion zone (Figure 3).

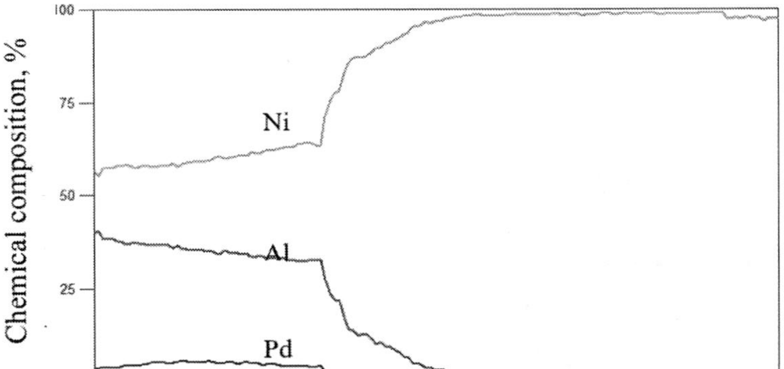

Figure 3. EDS analysis on the cross-section of the palladium modified aluminide coating deposited on the nickel substrate.

3.2 Palladium and hafnium modified aluminide coating on pure nickel

Palladium and hafnium modified aluminide coating on pure nickel has a double-zone structure (Fig. 4, Table 3). In the first, additive zone, on the top of the coating, the proportion of Ni to Al corresponds to the β-NiAl phase (Table 3, Point 1). Moreover, some palladium (10.2 at. %) and hafnium (0.3 at. %)

was identified. This indicates formation of the hafnium doped β-(Ni,Pd)Al phase. The thickness of this zone is ~45 μm. The second, interdiffusion zone, situated below is much thinner (~3 μm). Its chemical composition suggests formation of the palladium and hafnium doped γ'-Ni_3Al phase (Table 3, Point 4). Bright, hafnium rich precipitates were observed in the additive zone (Fig. 4-5). EDS spectrum of particle 1 and 2 shows presence of aluminum, nickel, palladium and hafnium peaks (Fig. 5 b, c). The nickel substrate/hafnium doped γ'-Ni_3Al as well as the hafnium doped γ'-Ni_3Al/hafnium doped β-(Ni,Pd)Al interfaces were relatively flat (Fig. 4). Kirkendall-like porosity was observed within the hafnium doped β-(Ni,Pd)Al phase near the hafnium doped γ'-Ni_3Al phase (Fig. 5).

Table 3. Chemical composition on the cross-section of the palladium and hafnium modified aluminide coating deposited on the nickel substrate (in at.%)

Spot	Chemical composition			
	Al	Ni	Pd	Hf
1	35.8	70.1	10.2	0.3
2	33.2	73.3	7.8	1.1
3	33.2	74.7	7.3	0.0
4	23.4	85.7	1.7	0.5
5	-	100.0	-	-

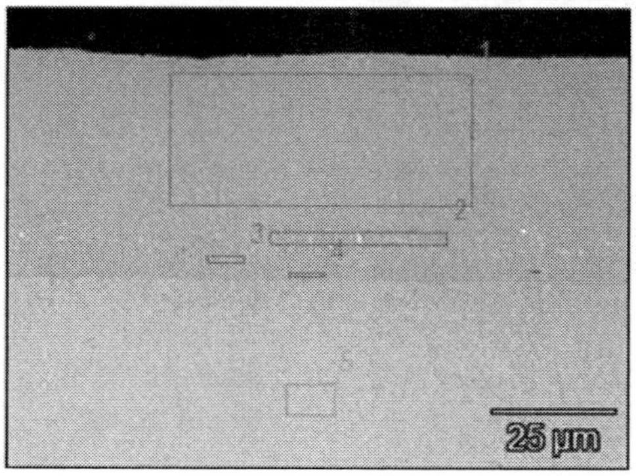

Figure 4. SEM microstructure on the cross-section of the palladium and hafnium modified aluminide coating deposited on the nickel substrate.

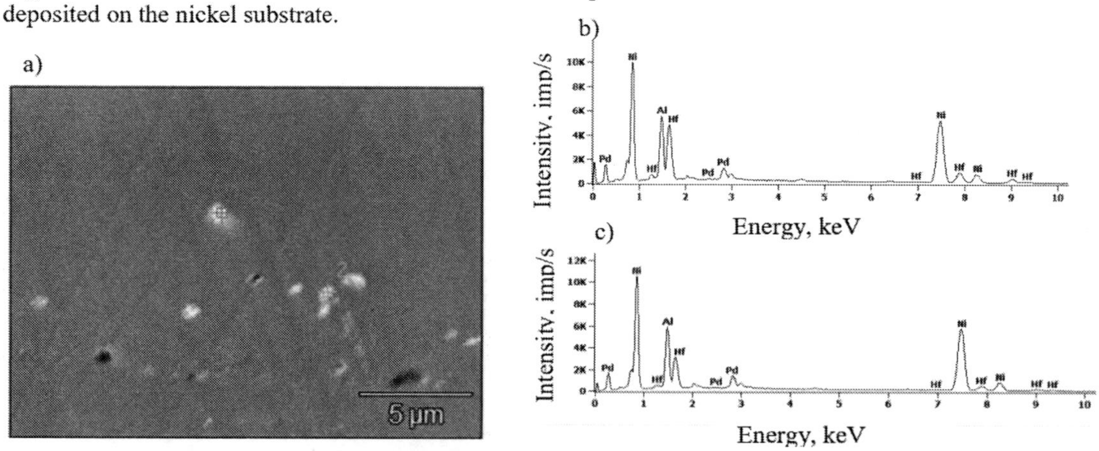

Figure 5. SEM microstructure of additive zone (a) with EDS spectrum of particle 1 (b) and particle 2 (c) of palladium and hafnium modified aluminide coating on the nickel substrate.

3.3 Palladium and zirconium modified aluminide coating on pure nickel

Palladium and zirconium modified aluminide coating on pure nickel has a double-zone structure (Fig. 6). In the first, additive zone, on the top of the coating, the proportion of Ni to Al corresponds to the β-

NiAl phase. Moreover, some palladium (3.7 at. %) was identified. This indicates formation of the β-(Ni,Pd)Al phase. The thickness of this zone is ~ 52 µm. The second, interdiffusion zone, situated below is much thinner (~6 µm). Its chemical composition suggests formation of the palladium doped γ'-Ni_3Al phase. Bright, zirconium rich precipitates were observed in the additive zone (Fig. 6 a, b, c, e). EDS spectrum of points 1 - 4 show presence of aluminum, nickel, palladium and zirconium peaks (Fig. 6 b - e). Kirkendall-like porosity was observed within the β-(Ni,Pd)Al phase near the γ'-Ni_3Al phase (Fig. 6).

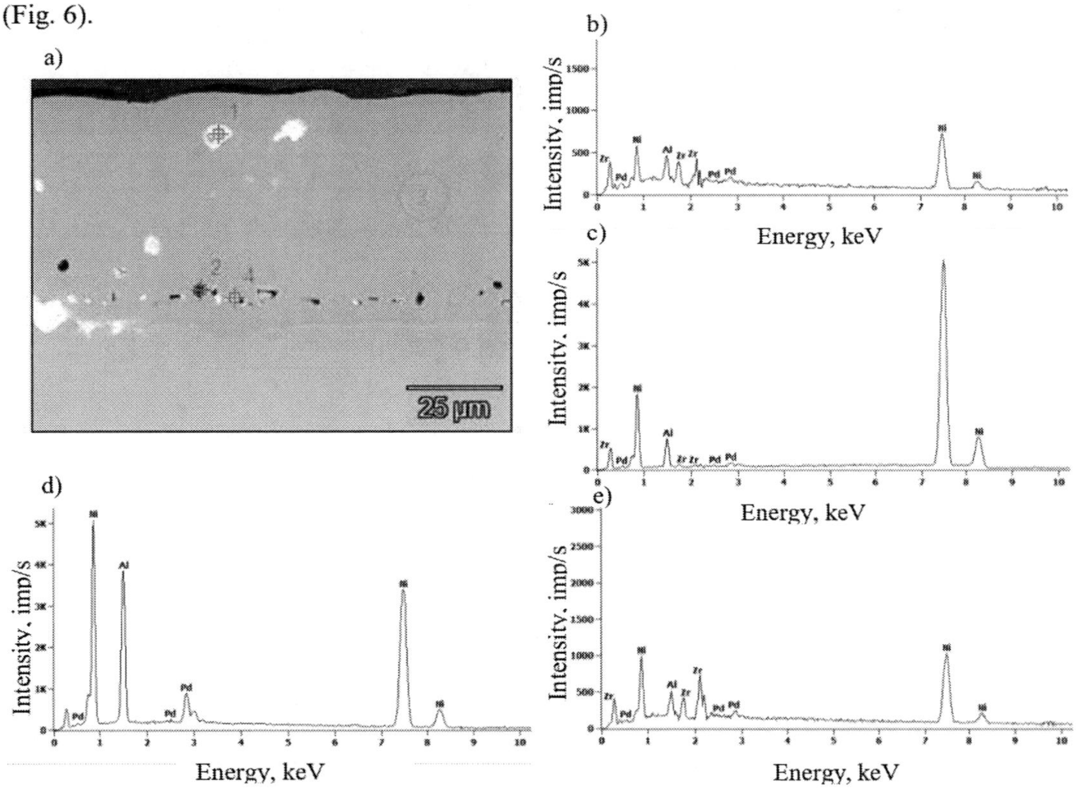

Figure 6. SEM microstructure on cross-section of palladium and zirconium modified aluminide coating deposited on the nickel substrate (a) with EDS spectrum of point 1-4 (b-e).

Palladium, palladium and hafnium and palladium and zirconium doped aluminide coatings show better oxidation resistance than the nonmodified ones (Fig. 7).

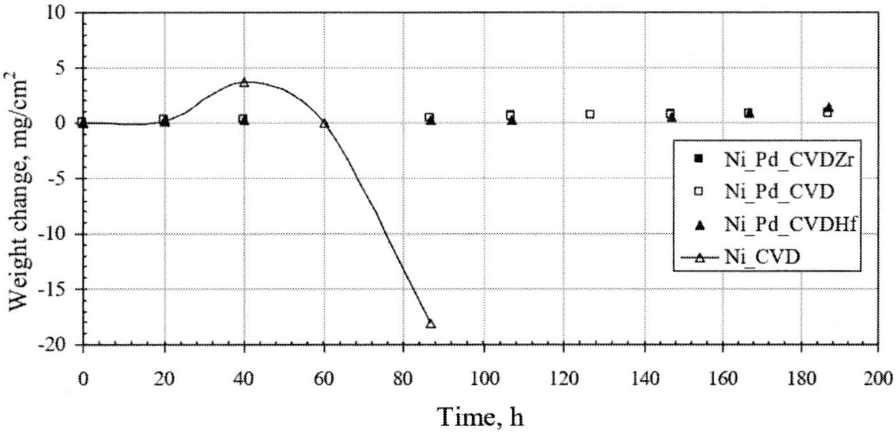

Figure 7. Weight change versus oxidation time coatings deposited on pure nickel.

4. Conclusions

The analyzed coatings consist of two zones, the additive and the interdiffusion ones. The additive zones in Pd modified and Pd+Zr modified coatings contain the the β-(Ni,Pd)Al phase, whereas in Pd+Hf modified one – the hafnium doped β-(Ni,Pd)Al phase The interdiffusion zone is formed by the palladium doped γ'-Ni_3Al or palladium and hafnium doped γ'-Ni_3Al phase phase. Palladium is distributed in whole coatings, whereas zirconium and hafnium is situated in precipitates in the additive zone. Oxidation resistance of co-doped Pd+Zr and Pd+Hf) coatings is significantly better than the single-doped (Pd) one. The sinegridtic effect of two dopants is obserwed.

Acknowledgment

The research has been supported by the National Science Centre, Poland Grant No. 2015/19/B/ST8/01645.

References

1. D.B. Miracle, Acta Metall. Mater. **41,** 649 (1993)
2. J.A. Haynes; K.L. More, B.A. Pint, I.G. Wright, K.M. Cooley, K.Y. Zhang, , High Temperature Corrosion and Protection of Materials. **5,** 35 (2000)
3. Y. Wang, G. Sayre, Surf Coat Technol, **203**, 1264 (2009)
4. S. Alperine, P. Steinmetz, P. Joso, A. Mater. Sci. Eng, A **121**, 367 (1989)
5. J.A. Haynes, Y. Zhang, K.M. Cooley, L. Walker, K.S. Reeves, B.A. Pint, Surf. Coat. Technol. **189**, 157 (2004)
6. D. Li, H. Guo, D. Wang, T. Zhang, S. Gong, X. Hubin, Corros Sci. **66**, 125 (2013).
7. S.J. Hong, G.H. Hwang, W.K. Han, S.G. Kang, Intermetallics **17, 381** (2009)
8. M. Zagula-Yavorska, J. Sieniawski, T. Gancarczyk, Arch. Met. and Mater. **57**, 503 (2012)
9. J. Romanowska, J. Morgiel, M. Zagula-Yavorska, J. Sieniawski, Materials Letters **145**, 162 (2015)
10. B. Ning, M. Shamsuzzoha, M.L. Weaver, Surf. Coat. Technol. **200**, 1270 (2005)
11. K.A. Unocic, B.A. Pint, Surf. Coat. Technol. **237**, 8 (2013)
12. D. Li, H. Guo, D. Wang, T. Zhang, S. Gong, X. Hubin, Corr. Sci. **66**, 125 (2013).
13. M. Zagula-Yavorska, J. Morgiel, J. Romanowska, J. Sieniawski, , Materials Letters **139,** 50 (2015)
14. S. Hamadi, M. Bacos, M. Poulain, A. Seyeux, V. Maurice, P. Marcus, Surf. Coat. Technol. **204**, 756 (2009)
15. R. Swadźba, M. Hetmańczyk, J.Wiederman, L. Swadźba, G. Moskal, B. Witala, K. Radwański, Surf Coat. Technol. **215**, 16 (2013)
16. B.A. Pint, K.L. More, I.G. Wright, Oxid Met, **59**, 257 (2003)
17. H. Guo, D. Li, L. Zheng, S. Gong, H. Xu, Corr. Sci. **88**, 197 (2014)
18. J.Romanowska, J. Morgiel, Ł.Kolek, P. Kwolek, M. Zagula-Yavorska, Arch. Civil Mech. Eng., **18**, 1421 (2018)
19. M. Zagula-Yavorska, J. Morgiel, J. Romanowska, J. Sieniawski, J. of Micr. **261**, 320 (2016)
20. J.Romanowska, M. Zagula-Yavorska, J. Sieniawski, Bul. Mater. Sci. **36,** 1043 (2013)

AUTHOR INDEX